Calculus and ODEs

Other titles in this series

Linear Algebra
R B J T Allenby

Mathematical Modelling
J Berry and K Houston

Discrete Mathematics
A Chetwynd and P Diggle

Particle Mechanics
C Collinson and T Roper

Ordinary Differential Equations
W Cox

Vectors in 2 or 3 Dimensions
A E Hirst

Numbers, Sequences and Series
K E Hirst

Groups
C R Jordan and D A Jordan

Probability
J McColl

In preparation

Analysis
E Kopp

Statistics
A Mayer and A M Sykes

Modular Mathematics Series

Calculus and ODEs

D Pearson

School of Mathematics
University of Hull

Elsevier Ltd.
Linacre House, Jordan Hill, Oxford OX2 8DP
200 Wheeler Road, Burlington, MA 01803

Transferred to digital printing 2004

© 1996 D Pearson

British Library Cataloguing in Publication Data
A catalogue record for this book is available from the British Library

ISBN 0 340 62530 9

Contents

Series Preface

This series is designed particularly, but not exclusively, for students reading degree programmes based on semester-long modules. Each text will cover the essential core of an area of mathematics and lay the foundation for further study in that area. Some texts may include more material than can be comfortably covered in a single module, the intention there being that the topics to be studied can be selected to meet the needs of the student. Historical contexts, real life situations, and linkages with other areas of mathematics and more advanced topics are included. Traditional worked examples and exercises are augmented by more open-ended exercises and tutorial problems suitable for group work or self-study. Where appropriate, the use of computer packages is encouraged. The first level texts assume only the A-level core curriculum.

Professor Chris D. Collinson
Dr Johnston Anderson
Mr Peter Holmes

Preface

This book has been written for those who have some experience of calculus at pre-university level, and who wish to develop their understanding of the subject to first year degree standard either in mathematics or in subjects involving a substantial amount of mathematics, with applications.

The emphasis throughout is both on the understanding and use of mathematical ideas. My aim has been to convey an appreciation of the fundamental ideas which underpin calculus, and even the more advanced student should find opportunities to acquire a deeper understanding of the subject. I have taken care not to make undue demands on the student's previous knowledge, and most topics are introduced from an elementary starting point, and without prior assumptions. Hand in hand with the development of a conceptual framework goes an emphasis on the methods and techniques which can be established as a consequence of the theory. Here the acquisition of manipulative skills is of the highest importance. The best way to learn mathematics is to do mathematics, and I hope that this will be made easier by the provision of more than 200 exercises throughout the book. In addition, I have set out numerous worked examples which will help to illustrate the ideas and methods of the subject.

I am indebted to Neil Gordon for his careful preparation of the manuscript, and to David Ross of Arnold for his help and patience. This book is dedicated to the three youngest mathematicians of my family, Paul, Alice and Christopher.

D. Pearson
School of Mathematics
University of Hull

1 • Introduction

The branch of mathematics known as calculus had its origins more than 300 years ago in the efforts of Newton, Leibniz, and others, to create a theory of motion, and to apply this theory to the science of dynamics. Calculus has evolved into a highly developed and efficient methodology for dealing with a variety of mathematical problems encountered in most areas where mathematics is used.

Underlying the modern theory is the idea of a function, more particularly a differentiable function, and the original goal of a theory of motion with applications to dynamics has been subsumed into the theory of differential equations. (Among Newton's countless other achievements, he was a pioneer in the early introduction of first order differential equations, which he referred to as fluxional equations, though this work was not published until long after his death.)

We can think of calculus, then, as a sophisticated tool for the analysis of functions, centred on the two notions of differentiation and integration, and having a bearing on such matters as the local variation and asymptotic behaviour of functions, graph sketching, the representation of functions by infinite series, maxima and minima, differential equations, and so on. Of course there have been progeny — whole areas of mathematics which have become independent areas of study in their own right, such as variational calculus, leading to the theory of optimization using in many cases additional methods drawn from other, quite separate branches of mathematics, and the major research area of functional analysis. It is common, nowadays, where the focus of interest is on formal aspects, mathematical structure, and rigour of argument rather than method and application, to refer to the subject as analysis, rather than calculus — so much so that it would sometimes appear that no self-respecting research mathematician would admit to a professional interest in calculus. There is, of course, no hard and fast dividing line between the two subjects, and it would be possible to regard this book as an introductory text in applied analysis.

1.1 Four aims

I have kept in mind, in writing this book, a number of aims which have guided me in the choice of content and style, and which correspond in my experience to the needs of students of calculus at first year university level. How far I have succeeded in meeting these aims will rest on the judgement of my readers.

The first aim was to provide, in reasonably compact form, an introduction to the properties of elementary functions, as revealed by the basic ideas and methods of differentiation and integration, and leading to a treatment of differential equations and related matters. I have assumed that the reader has studied mathematics to A-level or equivalent standard, and wishes to extend and develop this knowledge to a level which will enable him/her to cope confidently with the demands of calculus to first year university or college standard either in mathematics or in courses involving a substantial amount of mathematics.

I have not been content merely with a treatment of technique and method which leaves the student in ignorance of the fundamental principles and ideas underlying the applications. The student who can find the two maxima and one minimum of the function $y = |x|/(1 + x^2)$ and realize that at one of these three points $\mathrm{d}y/\mathrm{d}x$ does not exist, who can refrain from attempting to evaluate the integral $\int_{-1}^{1}(1/x^2)\mathrm{d}x$, or who can understand that the graph of the function $\sqrt{1 - e^{x^2}}$ may not be all that it appears, will in my view have a greater knowledge of calculus than the student who can do none of these things, but can nevertheless evaluate a range of complicated integrals. Both an understanding of the ideas and principles of the subject and the development of manipulative skills are important, and I have tried to cater for both; but technical skills mean little without understanding.

At the same time, the reader will not find in these pages detailed and rigorous proofs of the important results regarding convergence of series, existence of limits, integrals and derivatives, or of uniqueness theorems for differential equations. A more formal treatment in these areas would take the reader too far into the domain of analysis for an introductory text of this nature. I have drawn the line, too, at a thorough treatment of real numbers and their properties, believing that a more intuitive understanding of the number system will suffice at this stage, to be followed in due course, and in the light of developing experience, by a more careful exposition, perhaps in the second semester, for those for whom mathematics is a principal subject.

My second aim in this book was to convince the reader of the importance of the need to **do** mathematics. One can read mathematics and attend lectures and tutorials in mathematics, but the motivated student can make real progress by **doing** mathematics. To facilitate this, I have provided numerous exercises throughout the book, both in each section and at the end of chapters. Exercises following sections of the book are, to a greater extent, intended to develop understanding, particularly by encouraging the student to explore for him/herself some of the ideas and methods of the section. They are often to some degree open-ended, and may profitably be made the subject of group activity, for example as tutorial problems. Despite a common stereotype of the mathematician as an individual aloof and isolated from the real world, mathematics thrives on personal contact and the exchange of ideas, and in addition to **doing** mathematics the student should learn to **communicate** mathematics.

The exercises at the end of each chapter are intended to cover the main ideas and methods of the chapter, and will help to test the student's mastery of the material covered. Solutions are provided at the end of the book.

My third aim was to give an account of some of the more important applications of calculus, and to make clear the relevance of the ideas of the subject to a number of areas of everyday life and experience. I do not altogether share the concerns of a student at my own university who complained, in a questionnaire, that I had included too few problems on the filling of baths by water, which is, of course 'the principal subject matter of calculus'. The reader who seeks the help of calculus to fill a bath may be disappointed; but I have tried to explain, for example, why the physical world is to a large extent described by second order differential equations, and I do believe that the reader who has made a serious attempt to master the contents of this book will find much that can be applied to other areas.

Anyone involved in the teaching of calculus at university level, and increasingly at pre-university level, must take into account access of students to modern computer technology. At Hull, our own first year students attend a module on computational mathematics, in which they learn to program in PASCAL and make use of a range of mathematical software, available by network, covering such applications as graph sketching, numerical solution of algebraic and differential equations, and so on. Both DERIVE and MATHEMATICA, each with a facility of algebraic differentiation and integration, are available on the network to some users. In addition, students have their own scientific calculators, in most cases with a graphics display. While taking account of developments in computer technology, I have not thought it useful or appropriate to link the text to any **specific** use of mathematical software. While having every hope and expectation that the reader will make full use of opportunities which were unavailable even a few years ago, I consider that to place special emphasis on a particular programming language or software package would limit both the potential readership and the usefulness of the book itself.

My fourth aim was to convey a sense of the unity and coherence of the subject. Calculus is **not** a disparate body of unconnected methods and ideas. It is a consistent approach to a variety of problems, particularly involving the local behaviour of functions, with its own philosophy and applications. The study of calculus is not best served by an approach which adopts an individual method for every problem, taken from a kind of mathematical toolbox, while failing to recognize the principal unifying themes of the subject.

This book has developed from a course of lectures given in various forms over the years to first year students of mathematics and other disciplines at the University of Hull. The current course is taught over a 12 week semester, and the book may be used either for independent study or as a companion to an introductory course of similar length. As such it contains more material than can reasonably be covered in a single semester, so some degree of selection, both in material and in choice of exercises, will need to be made.

I have learnt a lot from successive groups of students of the course. I hope that they too have learnt from me, and will be satisfied with the outcome.

1.2 Learning about calculus

In the last section I rather emphasized my own role, or how I see it, in **teaching** calculus. Now I want to say something about **learning** calculus, and in particular about learning calculus from this book.

Each chapter deals with one of the major themes of calculus, and within the chapter each section explores some aspect of this theme, in a relatively self-contained way. The section is followed by a number of exercises which will help to develop understanding of the ideas of the section. The reader is advised to follow each section closely, noting particularly any special notations or definitions that are introduced, as well as taking account of the main ideas of the section. From time to time new **methods** will be introduced, usually accompanied by examples of applications of the method. These examples should be followed through carefully, while recognizing that an attempt to master the ideas and principles should always accompany putting these ideas and principles into practice.

The student of calculus will not usually have met the subject for the first time in this book. Some prior experience of calculus will help to reinforce some of the ideas introduced, and though in most instances I have preferred to assume the minimum background necessary, I would like to think that students make full use of any knowledge previously acquired. At the same time, the subject is developed here as a **university** discipline, that is in a treatment capable of supporting quite advanced applications and leading to further courses in later years. For this reason alone, it is necessary to adopt a rather more fundamental approach to some of the basic ideas. I hope the student will bear with me if he or she does not see every topic introduced in exactly the same way as in other texts, or with exactly the same kind of examples. I would ask the reader to take the trouble to look at the subject in a new light, and to take some time thinking through any new approach. The effort is well worth the trouble. Perhaps to some degree, particularly in the subject of calculus, we all need to free ourselves from old patterns of thought and realize once again the fundamental and even revolutionary aspects of the subject.

At the end of each chapter there are a number of exercises, usually of a more routine nature, which will serve as revision problems and test overall knowledge of the main theme treated in the chapter. I have provided fairly complete solutions to these exercises at the back of the book.

There are two major resources of which I would wish students to make full use while studying calculus with the help of this book. The first is the college, university or public library. Unless you think that this book (or your lecturer) has said the last word on the subject (and I do not think such a claim can reasonably be made for **any** book on mathematics, let alone any calculus text), you will wish to supplement your knowledge by appealing to other references on this and related subjects. It is of the greatest importance to consult other texts, both to see the subject presented in different contexts and with different notations, and also to see how calculus relates to other branches of mathematics and other areas of application. Mathematics is not, in the end, a group of unconnected, compartmentalized, subdisciplines, and it is quite possible, for example, to find that an idea you have come across in algebra or geometry may be of some relevance to calculus too!

The second resource of which I hope you, the student, will take advantage is that of the nearest computer laboratory to which you have access. Information technology is making formidable inroads into the teaching and learning environment, and I hope you will make full use of this. Among the packages currently available which will help you to carry out routine operations of calculus are DERIVE 3.0, MATHEMATICA 2.2, and MACSYMA 2.0. Other facilities may be available on your local network. Even the relatively humble scientific, programmable, or graphics calculator will be found to be a useful tool for many of the exercises in this book, with the usual proviso that these tools be regarded as an **aid** to understanding, rather than a **replacement** for understanding. In sketching the graph of a function, for example, the calculator should be the last rather than the first resort, and as such will often throw useful light on the nature of the function to be sketched, when used with proper care. (You will find I have thrown the odd spanner in the works, with some exercises in sketching which will prove a stiff test for most programs!)

2 • Functions

The subject matter of calculus is functions. In fact calculus could be described as that branch of mathematics which deals with functions and their local behaviour — we shall have more to say about that in Chapter 4. Progress in calculus must therefore be founded on a clear understanding of what functions are, and how they can be described and used. We shall also need to agree on some of the notations which will be used throughout this book.

2.1 What is a function?

The central idea of a function which you should have at the back of your mind is that of a rule, or prescription, for sending numbers to numbers. This is a good starting point, and we shall gradually refine and clarify the concept of function as we proceed.

'Numbers', here, will be taken to mean 'real numbers', rather than, say, complex numbers or rational numbers, or any of the other kinds of numbers that mathematicians have defined from time to time. It is possible to consider much more general notions of function in which we may want to consider rules for sending vectors to vectors, or matrices to numbers, or cabbages to kings, or just about anything to anything else; but, for the moment, let us agree on a function being defined by **a rule for sending real numbers to real numbers**.

◈ Example 1

Here is a simple example of such a function. The rule is: 'square the real number and then add two'.

This function is defined by the successive application of two operations (or functions), namely 'square' and 'add two', which must be carried out in the right order. If each of the two basic operations is represented by a box, the function 'square the real number and then add two' looks like the diagram in Fig 2.1; so this function sends 1 to 3, -5 to 27, $\sqrt{2}$ to 4, π to $\pi^2 + 2$, and so on.

Notice that the function is completely defined by the rule. There is no absolute need to introduce any additional notation at all in defining this function — we do not **need** to say that the function is y, or f, or g, or whatever, or that the input number is x, or t, or anything else. It is enough to give a clear and precise rule for sending numbers to numbers, and we then know what function we are talking

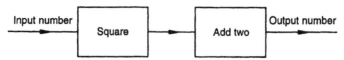

Fig 2.1   The function 'square and add two'.

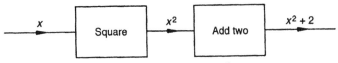

Fig 2.2 The function $x \rightarrow x^2 + 2$.

about; but let us not turn our backs on 1000 and more years of development of algebraic notation in mathematics!

Calling the input number x, and the output number y, our function is defined by the rule $x \rightarrow x^2 + 2$, and with $y = x^2 + 2$ the diagram of the function looks like Fig 2.2.

Of course, if we use the symbol t for the input number and θ for the output number, then the **same function** is given by the rule $t \rightarrow t^2 + 2$, and we can write $\theta = t^2 + 2$. Which notation we use will often depend on the context or on the particular applications we have in mind. You should try not to get too tied down by one particular notation, and get used to a variety of different ways of describing functions algebraically. A very common notation is to use the symbol f for the function (or rule) itself. In that case, for a given input number x, we denote by $f(x)$ the corresponding output number.

In the above example, input x and output y are related by $y = f(x) = x^2 + 2$. Often, instead of talking about the function f defined by the rule $f(x) = x^2 + 2$, we shall simply say 'the function $x^2 + 2$', or even 'the function $f(x)$'. Again, remember that other notations are possible, for example we may have 'the function $\theta(t)$ defined by $\theta(t) = t^2 + 2$'. Common symbols for functions in general are f, g, u, v, etc.

The following example illustrates the important point that a function $f(x)$ need not always be defined for **all** real numbers x; but whenever $f(x)$ **is** defined, there must be just **one** value of $f(x)$, not more, for given x.

Example 2

Define a function $f(x)$ by $f(x) = \sqrt{x^2 - 4}$. Here, and elsewhere in this book, we shall understand the **positive** square root by the symbol $\sqrt{\bullet}$ (with, of course, $\sqrt{0} = 0$). If, as for example in solving quadratic equations, we wish to indicate the two possibilities of positive/negative roots, this will be explicitly written $\pm\sqrt{\bullet}$.

The function $\sqrt{x^2 - 4}$ may be evaluated by a sequence of three successive operations, as shown in Fig 2.3. Note, however, that the third of these three operations cannot be carried out for any input value x such that $x^2 - 4 < 0$. This is because we cannot take the square root of a negative number. (A **complex** square root of a negative number can be defined; however, here we are dealing with

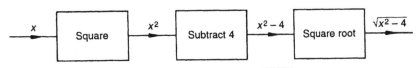

Fig 2.3 The function $x \rightarrow \sqrt{x^2 - 4}$.

functions in the context of **real numbers** only, and both input x and output y must be real.)

Since $x^2 - 4 < 0$ whenever $-2 < x < 2$, in this example $f(x)$ is undefined for x in the interval $-2 < x < 2$. The set of all (real) values of x for which $f(x)$ is defined is called the **domain** of the function $f(x)$. Since, in specifying precisely the rule $x \to f(x)$ by which a function is defined, it is necessary to state clearly the set of x to which this rule applies, it is good practice at least to make a mental note of the domain of each function that you want to define. Strictly, a function is not properly defined at all until you can say what its domain consists of. This usually causes little difficulty if you remember that the domain of a function f consists of all values of x for which the rule $y = f(x)$ makes sense.

In the case $f(x) = \sqrt{x^2 - 4}$, then, the domain is given by

$$\text{domain}(\sqrt{x^2 - 4}) = \{x; -\infty < x \le -2, \text{or } 2 \le x < \infty\} \tag{2.1}$$

or, using interval notation, with the set theoretic symbol $\cup$ for union,

$$\text{domain}(\sqrt{x^2 - 4}) = (-\infty, -2] \cup [2, \infty) \tag{2.2}$$

Here, square brackets], [mean that $x = -2, x = 2$ **are** to be included in their respective intervals, whereas round brackets (,) mean that $x = -\infty$, $x = +\infty$ (not being numbers at all!) are **not** included.

If we introduce the symbol $\mathbb{R}$ to denote the set of all real numbers, and denote by $A \backslash B$ the set of all x which are in A but not in B, an alternative notation is to write

$$\text{domain}(\sqrt{x^2 - 4}) = \mathbb{R} \backslash (-2, 2) \tag{2.3}$$

If we let x run through the domain of a function f, and set $y = f(x)$, then y will run through a set of real numbers called the **range** of the function f. Thus, with $y = f(x)$, remember that the **domain** is the set of all possible x values, whereas the **range** is the set of all possible y values. We insist, always, that each value of x in the domain of a function $f(x)$ gives rise to **one and only one** value of y. Such an expression as $\pm\sqrt{x^2 - 4}$, though making sense mathematically in the right context, **does not** define a function, since $y = f(x)$ must specify a unique value of y, for given x. In the above example, with $f(x) = \sqrt{x^2 - 4}$, the range contains no negative numbers at all, and

$$\text{range}(\sqrt{x^2 - 4}) = [0, \infty) \tag{2.4}$$

Most of the functions which we shall meet in this book are either available on standard 'scientific' calculators and computers, or are simply expressible as combinations of such functions. To 'enter' (input) a number into our calculator and observe the 'display' (output) when we press the square root 'function' button will make us appreciate the role of a function in defining a relationship between input and output numbers. The input of a negative number, followed by an 'error' display, should help us to understand the importance of domain, in the case of the square root function; there is, moreover, just one output number if we do not have an error message — our functions are all **single-valued**. While recognizing the value of the scientific calculator in making a range of functions accessible to us, we

Table 2.1 Some standard functions.

Function	Domain
x^2	$\mathbb{R}$
$1/x$	$\mathbb{R}\backslash\{0\}$
$|x|$	$\mathbb{R}$
$\sqrt{x}$	$[0, \infty)$
$\sqrt[3]{x}$	$\mathbb{R}$
$\sin x$	$\mathbb{R}$
$\exp(x)$	$\mathbb{R}$
$\ln(x)$	$(0, \infty)$
$\log_{10}(x)$	$(0, \infty)$

should not overlook its limitations. No calculator can give us the 'exact' value of the square root of 2 — if interpreted literally (numerically?) my calculator's evaluation of $\sqrt{2}$ to eight decimal places tells me (incorrectly) that $\sqrt{2}$ is rational; and an error message on evaluating π raised to the power 250 tells us more about the capacity of our calculator than about the domain of any function!

Table 2.1 lists some of the functions available from a standard scientific calculator, together with their respective domains. Do not worry if you are unsure of the precise definitions and properties of some of these functions. Any that you are uncertain about will be considered in more detail later in the book. The third function on the list, $|x|$, is the modulus function (pronounced 'mod. x') or absolute value function, and is defined by

$$|x| = x, \quad \text{for } x \geq 0$$
$$= -x, \text{for } x < 0 \tag{2.5}$$

In the case of the fifth function in the list, the cube root function $\sqrt[3]{x}$, note that the domain is $\mathbb{R}$ (you **can** take the cube root of negative numbers — in fact $\sqrt[3]{-x} = -\sqrt[3]{x}$). However, a calculator evaluation of $\sqrt[3]{x}$ for $x < 0$ may possibly lead to an error message because $\sqrt[3]{x}$ is interpreted by the calculator as $\exp(\ln(x)/3)$, and negative numbers are not in the domain of $\ln(x)$; see Chapter 7 for the definition of $\ln(x)$ and the related question of definition of fractional powers.

EXERCISES ON 2.1

1. Make a list of the functions available to you from your calculator (or computer). For those functions which are familiar to you, state what is their domain and range, and explain as far as you can how these functions are defined. (You will gradually be able to develop and refine your answer to this question as you progress through the book.)
2. Define some simple functions of your own which can be obtained through a sequence of two or three operations. Try to determine the domain and range in each case.

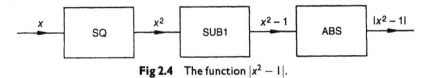

Fig 2.4   The function $|x^2 - 1|$.

3. Three operations, called respectively SQ, SUB1, ABS, are defined as follows:
   SQ is the operation, or function, which sends a number to its square;
   SUB1 subtracts one from the input number;
   ABS takes the absolute value, or modulus.
   Figure 2.4 represents the function $|x^2 - 1|$ as resulting from the successive application of the three operations SQ, SUB1, ABS, in that order. Construct similar diagrams for the functions representing the successive application of the three operations in each of the remaining five possible orderings (SQ, ABS, SUB1), (SUB1, SQ, ABS), etc. Write down the function $f(x)$ obtained in each case, and verify that the six possible orderings produce in total only four distinct functions.

2.2 Functions and their graphs

The graph of a function $f(x)$ is simply a plot of y against x, using two-dimensional (rectangular, Cartesian) coordinates (x, y), where y and x are related by $y = f(x)$. Often, in trying to understand the behaviour of a function $f(x)$, it is enough to carry out a rough sketch of the graph, indicating overall features such as intercepts with the coordinate axes, asymptotes, points of maxima/minima, and so on. I shall have much to say about various aspects of graph sketching in later chapters. See in particular Chapter 6.

You will certainly be acquainted already with the graphs of a number of functions. Figures 2.5 and 2.6 are examples of graphs for a few of the functions which we have met so far in this chapter. These graphs illustrate some of the ideas we have met in this chapter. Note first of all that, in order to qualify to be the graph of a function, any vertical line must meet the graph in **at most** one point. (Why?) For example, a (complete) circle could not be the graph of any function, since a vertical line can meet the circle in two points.

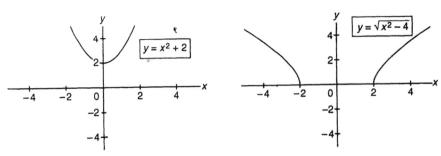

Fig 2.5   Graphs of $y = x^2 + 2$ and $y = \sqrt{x^2 - 4}$.

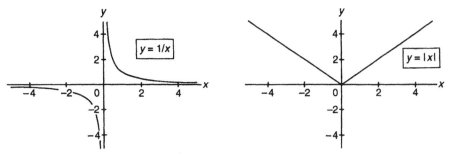

Fig 2.6 Graphs of $y = 1/x$ and $y = |x|$.

A given real number a will belong to the **domain** of $f(x)$ provided the vertical line $x = a$ meets the graph, i.e. contains one point of the graph. If, on the other hand, $x = a$ does not intersect the graph, then a does not belong to this domain. Applying these ideas to the second graph in Fig 2.5, for example, confirms our previous conclusion that domain$(\sqrt{x^2 - 4}) = \mathbb{R}\backslash(-2, 2)$. A given real number b will belong to the **range** of $f(x)$ provided the horizontal line $y = b$ meets the graph; if this line does meet the graph it may do so in one or more points, depending on the function.

The modulus function (second graph in Fig 2.6) is about the simplest example of a function for which $f(x)$ is defined differently for x belonging to different subsets of $\mathbb{R}$. Thus, according to equation (2.5), we have $|x| = +x$ for $x \in [0, \infty)$ ($\in$ means 'belongs to'), whereas $|x| = -x$ for $x \in (-\infty, 0)$. Functions involving the modulus are often best interpreted by 'translating' the modulus, whenever it occurs, into its value on the various subintervals for which the function is defined. The following example illustrates the sketching of graphs in such cases.

✴ *Example 3*

It is required to sketch graphs for each of the functions $|x(1 - x)|$, $x(1 - |x|)$, and $|x + 1| + |x - 1|$.

In the first case, interpret the function as

$$f(x) = x(1 - x), \text{ whenever } x(1 - x) \geq 0$$
$$= -x(1 - x), \text{ whenever } x(1 - x) < 0$$

Since $x(1 - x) \geq 0$ whenever $0 \leq x \leq 1$, we must have $y = x(1 - x)$ for x in this interval and $y = -x(1 - x)$ outside this interval. Note that the graph $y = -x(1 - x)$ is the reflection in the x-axis (i.e. $y \to -y$) of the graph $y = x(1 - x)$. Hence $y = |x(1 - x)|$ is the continuous curve in Fig 2.7.

Similar arguments apply to the second and third functions in this example. Note

$$x(1 - |x|) = x(1 - x), \text{ for } x \geq 0$$
$$= x(1 + x), \text{ for } x < 0.$$

In the case $f(x) = |x + 1| + |x - 1|$, observe that crucial changes of sign occur around each of the points $x = -1$ and $x = +1$. Hence we have **three** intervals to

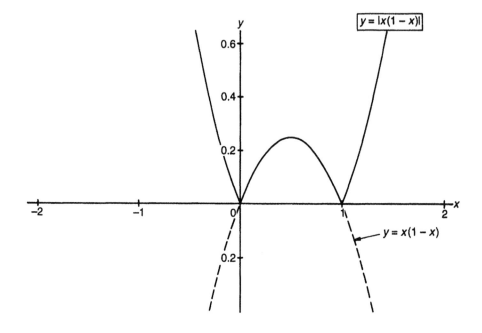

Fig 2.7 The graph of $y = |x(1-x)|$.

consider, namely $x < -1$, $-1 \leq x < 1$, and $x \geq 1$. 'Translating' the modulus in each of these three regions, we then have

$$f(x) = -(x+1) - (x-1) = -2x, x \in (-\infty, -1)$$
$$= (x+1) - (x-1) = 2, x \in [-1, 1)$$
$$= (x+1) + (x-1) = 2x, x \in [1, \infty).$$

These two remaining graphs are shown in Fig 2.8.

Many other examples of functions defined differently on different subsets of $\mathbb{R}$ occur frequently and are important in applications. In Example 4 below, we

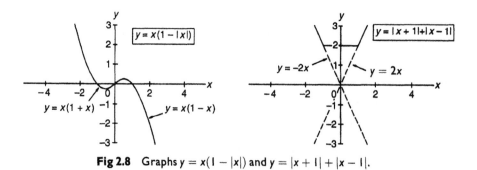

Fig 2.8 Graphs $y = x(1 - |x|)$ and $y = |x+1| + |x-1|$.

consider the graphs of two such functions. The first is the integer part function, defined by

$$\text{INT}(x) = n, \quad \text{for } x \in [n, n+1)$$

for any integer n (positive, negative, or zero). Thus $\text{INT}(\sqrt{2}) = 1$, $\text{INT}(\pi) = 3$, $\text{INT}(-3.2) = -4$.

The second function $\psi(t)$ shows that there need be no relation whatever between the values taken by the function on the respective subintervals of its domain.

⊕ *Example 4*

It is required to sketch the graphs of the functions $\text{INT}(x)$ and $\psi(t)$, where $\psi(t)$ is defined by

$$\psi(t) = t + 1, t \in (-\infty, 0)$$
$$= t^2, t \in [0, \infty)$$

These graphs are given in Fig 2.9. Note, for example, that $\text{INT}(0) = 0$ and $\psi(0) = 0$. Where, in a graph, there is possible ambiguity, we indicate by a closed circle a point that **is** on the graph, and by an open circle a point that **is not**.

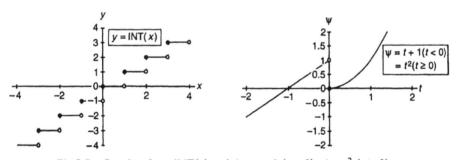

Fig 2.9 Graphs of $y = \text{INT}(x)$ and $\psi = t + 1$ $(t < 0)$, $\psi = t^2$ $(t \geq 0)$.

EXERCISES ON 2.2

1. Sketch, in a single diagram, the graphs of the functions x, x^2, x^3 and x^4. What does the graph of x^N look like, if N is a very large positive integer? Sketch, on another diagram, the graphs of the functions x, $\sqrt{x}$, $\sqrt[3]{x}$ and $\sqrt[4]{x}$, for $x \geq 0$. What does the graph of $\sqrt[N]{x}$ look like, if N is a very large positive integer?

2. Fig 2.10 shows the graph of a function $f(x)$. Note that $f(0) = f(1) = 0$, and that the function has a maximum value of $+1$ and a minimum value of -1. Sketch the graph of each of the functions: $2f(x)$, $f(x) + 1$, $f(x + 1)$, $f(x - 1)$, $f(2x)$, $f(-x)$, $f(|x|)$ and $|f(x)|$. State the **range** of each of these functions.

3. Sketch the graph of each of the four functions which you obtained in question 3 of Exercises on 2.1.

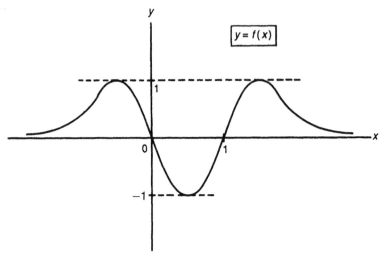

Fig 2.10 The graph of a function $f(x)$.

2.3 What is a continuous function?

This book is about calculus. It is not a book on analysis. I assume then, that you would not expect from this book a detailed treatment of notions such as continuity, limit and so on, with formal definitions and rigorous proofs. That approach to the subject is very much the concern of analysis. I hope that many of my readers will wish to continue their studies into analysis, perhaps after completing this book. Certainly their needs are well catered for by a further volume in this series, and it is one of my aims to have whetted appetites for future work. At my own university, we run a one semester module in calculus in parallel with a module on mathematical reasoning, to be followed by a module on analysis. This does **not** mean, however, that calculus is simply analysis without the reasoning, without the rigour, and without the need for clear thinking, as it has sometimes been caricatured. I feel that I would be failing in my duty to the reader if I did not attempt to give some idea of what the notions of continuity and limit mean and why they can be important even to the student who does not intend to pursue the stricter and narrower demands of rigour and abstraction. I shall steer clear of the usual ϵ, δ definition of continuity, and instead present an approach which will, I hope, get over the main ideas while being mathematically correct and leading to useful applications.

What is a continuous function? Let f be a function and x_0 be a point of the domain of f. To start with, we should be clear what it means for the function f to be continuous at x_0. What this means, very roughly, is that if x is close to x_0 then $f(x)$ is close to $f(x_0)$, and that as x approaches x_0 then $f(x)$ will approach $f(x_0)$. This is, however, too imprecise to qualify as a definition of continuity. We can do a little better by introducing the notion of the **variation** of a function in an interval. Let δ be a positive number, and assume that not only x_0 but **all** points of the interval $x_0 - \delta \leq x \leq x_0 + \delta$ belong to the domain of the function f. Let us denote by $V(\delta)$ the **variation** in the function $f(x)$, for x in the interval $[x_0 - \delta, x_0 + \delta]$. As a first

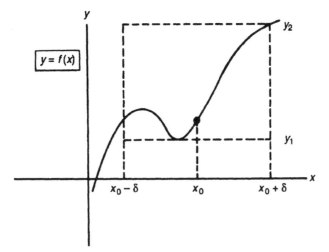

Fig 2.11 Variation of a function on an interval.

orientation to this idea, you can think of $V(\delta)$ as being the difference between the largest and the smallest values of $f(x)$, for x in the given interval; that is, the variation measures how much the function's value changes as x runs across the interval. Though this way of thinking of $V(\delta)$ is perfectly acceptable and correct if f **has** a largest and a smallest value in the interval, it may be that no such maximum/minimum values exist, so a better and generally applicable definition of $V(\delta)$ is to say $V(\delta)$ is the length $(y_2 - y_1)$ of the shortest interval $[y_1, y_2]$ such that the values of $f(x)$, for $x \in [x_0 - \delta, x_0 + \delta]$, always lie in the range $[y_1, y_2]$. With this definition of variation, $V(\delta)$ will always exist, with the convention that $V(\delta) = \infty$ if the function is so wild that there is **no** finite interval $[y_1, y_2]$ big enough to contain the values of $f(x)$, and at the other extreme $V(\delta) = 0$ if f is a constant function, for $x \in [x_0 - \delta, x_0 + \delta]$.

Figure 2.11 explains how to give a graphical interpretation to the variation $V(\delta)$. For x in the interval $[x_0 - \delta, x_0 + \delta]$, with $y = f(x)$, we must have $y_1 \le y \le y_2$, and $[y_1, y_2]$ must be the **shortest** interval having this property. This means that y_1 cannot be any larger, nor can y_2 be any smaller. For $x \in [x_0 - \delta, x_0 + \delta]$, we have put the graph $y = f(x)$ inside a rectangle having sides $x = x_0 - \delta$, $x = x_0 + \delta$, $y = y_1$ and $y = y_2$. Given that the vertical sides of the rectangle are already prescribed, to fit with the endpoints $x_0 - \delta$, $x_0 + \delta$ of the given interval, our rectangle is the smallest such rectangle into which the graph will fit. The width of the rectangle is 2δ, and the height of the rectangle, $(y_2 - y_1)$, is what we call the variation $V(\delta)$ of the function f over the interval $[x_0 - \delta, x_0 + \delta]$. Provided that f has a maximum and minimum value on this interval, this variation will come out as just $f_{max} - f_{min}$.

The idea of continuity for the function f at $x = x_0$ is bound up with what happens to the variation $V(\delta)$ as δ is reduced and approaches zero. The first and obvious thing to notice is that with $0 < \delta_1 < \delta_2$ we must have $V(\delta_1) \le V(\delta_2)$. This is because any interval $[y_1, y_2]$ such that $y_1 \le f(x) \le y_2$ for $x_0 - \delta_2 \le x \le x_0 + \delta_2$ must also satisfy $y_1 \le f(x) \le y_2$ for $x_0 - \delta_1 \le x \le x_0 + \delta_1$. A function cannot have greater variation on the smaller interval than it has on the larger! As the value of δ

is steadily reduced, say from $\delta = 1$ with $0 < \delta \leq 1$, the variation $V(\delta)$ can only get smaller or remain constant — it cannot increase. There are only two possibilities. These are either:

(a) $V(\delta)$ stays strictly away from zero ($V(\delta) \geq$ const > 0) for all δ in the interval $0 < \delta \leq 1$; or
(b) $V(\delta)$ reduces to zero as δ reduces to zero — there is then **no** positive constant such that $V(\delta) \geq$ const > 0 for all δ in the interval $0 < \delta \leq 1$.

If (a) holds, we shall say that the function f is **discontinuous** at $x = x_0$; if (b) holds, we shall say that f is **continuous** at $x = x_0$; so f will be continuous at x_0 whenever the variation in f on the interval $[x_0 - \delta, x_0 + \delta]$ goes to zero as δ goes to zero, and discontinuous at x_0 if the variation stays strictly positive, however small the value of δ becomes. In terms of the graphical interpretation of Fig 2.11, continuity at x_0 means that the **height** $(y_2 - y_1)$ of the enclosing rectangle goes to zero as the width, 2δ, goes to zero. If f is discontinuous at x_0, on the other hand, the height $(y_2 - y_1)$ can never reduce below a certain strictly positive value H, which is a measure of the jump or variation in height of the graph near $x = x_0$.

Notice that, for a function to be continuous at x_0, we do demand (among other things) that at least if δ is small enough the interval $[x_0 - \delta, x_0 + \delta]$ should belong entirely to the domain of f. Otherwise, again we shall say that f is discontinuous at x_0. A function continuous at x_0 must certainly then be defined for all x near enough to x_0 — a function of which the domain is, say, only a single point in an interval must be classified as discontinuous.

A function f which is continuous at **every** point of its domain is called, simply, a continuous function. Any other function is called discontinuous, so if there is just one (or more) point at which a function is discontinuous, then that function is called a discontinuous function. The rules for belonging to the company of continuous functions are quite exclusive and strictly adhered to. On the other hand, the benefits of membership are considerable.

⊛ *Example 5*

Two functions which illustrate some of the above ideas in a simple way are the absolute value function $f(x) = |x|$, and the so-called Heaviside step function $h(x)$, which is defined by

$$h(x) = 1 \quad (x \geq 0)$$
$$\quad\;\; = 0 \quad (x < 0).$$

To consider first of all the absolute value function $f(x) = |x|$, let x_0 be any positive number. Then if δ is small enough, $\delta < x_0$, the interval $[x_0 - \delta, x_0 + \delta]$ contains only positive numbers, for which $f(x) = x$. On the interval, $f(x)$ ranges between its minimum value of $x_0 - \delta$ and its maximum value of $x_0 + \delta$, so the shortest interval $[y_1, y_2]$ such that $y_1 \leq y \leq y_2$ for $y = f(x)$ is the interval $[x_0 - \delta, x_0 + \delta]$ itself, which has length $V(\delta) = 2\delta$. Clearly, 2δ goes to zero as δ goes to zero, so that f is continuous at x_0. Similarly, f is continuous at any negative number x_0.

If we take $x_0 = 0$, the function $|x|$ for $-\delta \leq x \leq \delta$ has minimum value zero (at $x = 0$) and maximum value δ (at $x = -\delta$ and at $x = +\delta$), so in this case we have $V(\delta) = \delta - 0 = \delta$, which again goes to zero as δ goes to zero. Thus $f(x) = |x|$ is also continuous at $x = 0$. Hence f is continuous at **all** points of its domain, and we can say that the absolute value function is a continuous function.

In the case of the Heaviside function $h(x)$, it is easy to see that any $x_0 \neq 0$ is a point of continuity. For $x_0 = 0$, the situation is different, since in any interval $[-\delta, \delta]$ with $\delta > 0$, h has a maximum value 1 and a minimum value 0. Hence $V(\delta) = 1 - 0 = 1$ in this case, and $V(\delta)$ certainly does **not** reduce to zero as δ approaches zero. The Heaviside function is therefore discontinuous at $x = 0$; we say that h has a jump discontinuity of magnitude 1 at the origin. Evidently h is a discontinuous function.

⊛ *Example 6*

Continuity of powers: it is not difficult to verify the continuity of $f(x) = x^N$ for any integer power N, whether positive or negative. As simple but typical examples, consider the functions $f_1(x) = x^2$ and $f_2(x) = x^{-1} = 1/x$. For the interval $[x_0 - \delta, x_0 + \delta]$ with $x_0 > 0$ and $0 < \delta < x_0$, $f_1(x)$ ranges in the first case between $(x_0 - \delta)^2$ and $(x_0 + \delta)^2$, so that $V(\delta) = (x_0 + \delta)^2 - (x_0 - \delta)^2 = 4x_0\delta$, with a similar result for $x_0 < 0$, and $V(\delta) = \delta^2$ for $x_0 = 0$. Since $V(\delta)$ reduces to zero, f_1 is continuous in this case.

For $f_2(x) = 1/x$, consider the interval $[x_0 - \delta, x_0 + \delta]$ with, for example, $0 < \delta < x_0/2$. Then

$$V(\delta) = \frac{1}{x_0 - \delta} - \frac{1}{x_0 + \delta} = \frac{2\delta}{x_0^2 - \delta^2} < \frac{2\delta}{x_0^2 - (x_0/2)^2} = \frac{8\delta}{3x_0^2}$$

which reduces to zero as δ goes to zero. Hence f_2 is continuous at any point $x_0 > 0$, and a similar argument shows the function to be continuous at any negative point. Since $x = 0$ is not in the domain, we have shown that $f_2(x)$ is continuous at every point of its domain. The reciprocal function is a continuous function.

Similar arguments apply to other integer values of N. As far as non-integer powers are concerned, a little care needs to be taken. The function $x^{1/3}$, for example, the cube root function, when defined for all $x \in \mathbb{R}$, **is** a continuous function. However, a function like $\sqrt{x}$ is not continuous at $x = 0$. This is because $x = 0$ cannot be put at the centre of an interval $[-\delta, \delta]$ of points all of which belong to the domain of the function. The point $x = 0$ is an endpoint of the interval $[0, \infty)$ which defines the domain of the square root function, and points on the edge of a domain cannot qualify as points of continuity. Thus the square root function, though continuous at any $x_0 > 0$, is not continuous at one of the points of its domain, and must therefore be classified as a discontinuous function.

⊛ *Example 7*

Continuity of trigonometric and related functions: the trigonometric functions $\sin x$, $\cos x$ and $\tan x$ are all continuous functions. I do not intend to verify this statement by means of a complete proof. However, the case $f(x) = \cos x$ is treated in the exercises at the end of this section, and similar arguments apply to $\sin x$ and $\tan x$.

We shall meet many other examples of continuous functions in this book. The exponential, logarithmic and related functions are all continuous. Any function that can be **differentiated** is continuous; in particular, solutions of differential equations are continuous functions. Why is it necessary to draw attention to this property of continuity? In the first place, it is good to know of a function that a small change in input value leads to a correspondingly small change in output value. Very few functions can be calculated with exact precision, and it would be disconcerting if we had to input exact values of x in order to achieve reliable values for $f(x)$.

We can also see the important role of continuity in many of the fundamental results of analysis. An example is the so-called Intermediate Value Theorem. This theorem states that if f is any continuous function defined on an interval $a \leq x \leq b$, such that $f(a) < 0$ and $f(b) > 0$, then there exists at least one point, say $x = c$, in the interval, at which $f(c) = 0$. The idea underlying the Intermediate Value Theorem is that a function continuous on an interval cannot jump from one value to another without assuming all values between. The continuous function $x^3 - x - 1$, for example, which is negative at $x = 1$ and positive at $x = 2$, must have the value zero at some intermediate point; in other words, the cubic equation $x^3 - x - 1 = 0$ has a root between 1 and 2.

Continuous functions play a vital role in modern mathematics. We shall not need to explore all the ramifications of the theory, but a little time taken to gain at least an intuitive idea of this far reaching concept will be time well spent.

EXERCISES ON 2.3

1. Continuous functions are sometimes characterized as those for which it is possible to draw the graph without lifting the pen from the paper. Discuss this characterization, with particular reference to examples of continuous and discontinuous functions, and explain its limitations.

2. Determine the variation $V(\delta)$ of the function $\sqrt{1 + x^2}$ on the interval $x_0 - \delta \leq x \leq x_0 + \delta$, for small positive values of δ, distinguishing between the cases $x_0 < 0$, $x_0 = 0$, and $x_0 > 0$. Hence show that $\sqrt{1 + x^2}$ is a continuous function. Why is $\sqrt{1 + x}$ **not** a continuous function?

3. Show that the variation of $|f(x)|$ on an interval cannot exceed the variation of $f(x)$ on the same interval; hence show that if f is a continuous function then so is $|f|$.

4. A function f satisfies the inequality $|f(x_1) - f(x_2)| \leq |x_1 - x_2|$, for all $x_1, x_2 \in \mathbb{R}$. Show that the variation $V(\delta)$ of f on the interval $x_0 - \delta \leq x \leq x_0 + \delta$ satisfies $V(\delta) \leq 2\delta$. Hence show that f is a continuous function.

   Generalize this result to show that if f and g are two functions such that $|f(x_1) - f(x_2)| \leq c|g(x_1) - g(x_2)|$ for all x_1, x_2 in the domain of g, where c is a positive constant, then if g is a continuous function, so is f.

5. Assuming the identity

$$\cos A - \cos B = 2 \sin\left(\frac{A + B}{2}\right) \sin\left(\frac{B - A}{2}\right)$$

together with the inequality $|\sin \theta| \leq |\theta|$, show that the variation $V(\delta)$ of the function $\cos x$ on the interval $x_0 - \delta \leq x \leq x_0 + \delta$ cannot exceed 2δ. Deduce that $\cos x$ is a continuous function.

2.4 What is a limit?

The idea of a limit is closely tied to the idea of continuity. Often, limits are defined first, and then continuous functions are defined using limits. On the other hand it is perfectly respectable, and we have adopted this procedure here, to look at continuity before going on to examine the idea of a limit.

The basic notion is that of a function $f(x)$ tending to a limit l as x approaches a real number x_0. We shall write this as $\lim_{x \to x_0} f(x) = l$. The actual value $f(x_0)$ of the function f at $x = x_0$ is not really relevant; indeed the point $x = x_0$ need not even be in the domain of f. You can think of l as the value that $f(x)$ gets closer and closer to, as x approaches x_0 — but you never actually have to put x **equal** to x_0.

Given a function $f(x)$, defined for all x near to some point x_0, $f(x)$ need not tend to a limit at all as x approaches x_0. In order to decide whether or not $f(x)$ does approach a limit as $x \to x_0$, again we can make use of the **variation** of the function in an interval. In this case, since the value of f at $x = x_0$ is irrelevant, we can exclude this point, and define $V_0(\delta)$ to be the variation in the value of $f(x)$, as x runs over the interval $[x_0 - \delta, x_0 + \delta]$, with x_0 itself excluded. That is, $V_0(\delta)$ is the length $(y_2 - y_1)$ of the smallest interval $[y_1, y_2]$ such that $y_1 \le f(x) \le y_2$ for all $x \in [x_0 - \delta, x_0 + \delta] \backslash \{x_0\}$ (equivalently, for all x in $[x_0 - \delta, x_0) \cup (x_0, x_0 + \delta]$). Hence $V_0(\delta)$ is a measure of how much the function changes, within distance δ of x_0, but taking no account of the value of f at x_0, if indeed $f(x_0)$ is defined. Again, as the value of δ is reduced towards zero, two possibilities can occur. Either $V_0(\delta) \ge \text{const} > 0$ for all δ in the interval $0 < \delta \le 1$ or $V_0(\delta)$ reduces to zero as δ is reduced to zero. Only if the second alternative holds can we define a limit for the function $f(x)$ as x approaches zero. If $V_0(\delta)$ goes to zero the two endpoints of the shortest interval $[y_1, y_2]$ such that $f(x) \in [y_1, y_2]$ for $x \in [x_0 - \delta, x_0 + \delta] \backslash \{x_0\}$ will approach each other. There will then be a unique value y_0 such that $y_0 \in [y_1, y_2]$ for all δ in the interval $0 < \delta \le 1$. This number y_0 can be defined to be the limit l; that is, we have $y_0 = \lim_{x \to x_0} f(x)$. If, on the other hand, $V_0(\delta)$ remains strictly away from zero, then no such limit exists.

Assuming that $y_0 = \lim_{x \to x_0} f(x)$ **does** exist, it is possible to add a further point x_0 to the domain of f by setting $f(x_0) = y_0$. Since $y_0 \in [y_1, y_2]$, this additional point in the domain will not change the variation of f in the interval $[x_0 - \delta, x_0 + \delta]$; in terms of the notation of Section 2.3, this means that $V(\delta) = V_0(\delta)$. It follows that $V(\delta)$ goes to zero as δ approaches zero. In other words, to set $f(x_0) = y_0$ makes $f(x)$ into a **continuous function** at $x = x_0$. Thus $l = \lim_{x \to x_0} f(x)$ is the unique real number such that, with $f(x_0) = l$, the function f is continuous at the point x_0.

❄ *Example 8*

Define a function ψ by $\psi(t) = t/(\sqrt{1 + t} - 1)$. This definition makes sense for all $t \ge -1$, except for $t = 0$, so the domain of ψ is $[-1, \infty) \backslash \{0\}$, or equivalently the set $[-1, 0) \cup (0, \infty)$. Observe that ψ is not a continuous function, since the point $t = -1$ in the domain of ψ is at the edge of the domain. We examine the limit of $\psi(t)$ as t tends to zero. To do so, consider the values of $\psi(t)$ for $t \in [-\delta, \delta]$, omitting however the point $t = 0$; we take δ to satisfy $0 < \delta \le 1$. The function ψ is actually another function in disguise! For all t in the domain of ψ, on multiplying numerator and denominator by $\sqrt{1+t}+1$, we find $\psi(t) = t(\sqrt{1+t}+1)/[(1+t)-1] = \sqrt{1+t}+1$.

For $t \in [-\delta, \delta]$, $t \neq 0$, $\psi(y)$ must lie between the values $y_1 = \sqrt{1-\delta}+1$, $y_2 = \sqrt{1+\delta}+1$, and the smallest interval such that $y_1 \leq \psi(t) \leq y_2$ has length $y_2 - y_1 = \sqrt{1+\delta} - \sqrt{1-\delta}$. Hence we find in this case $V_0(\delta) = \sqrt{1+\delta} - \sqrt{1-\delta} = (\sqrt{1+\delta} - \sqrt{1-\delta})/1$, which on multiplying numerator and denominator by $\sqrt{1+\delta} + \sqrt{1-\delta}$ becomes $V_0(\delta) = 2\delta/(\sqrt{1+\delta} + \sqrt{1-\delta})$. Certainly, for δ between 0 and 1, $\sqrt{1+\delta} > 1$ and $\sqrt{1-\delta} \geq 0$, so that $V_0(\delta) < 2\delta$. Hence $V_0(\delta)$ reduces to zero as δ goes to zero. We conclude that $\psi(t)$ **does** have a limit as t tends to zero. To determine this limit, we need only note that $y_0 = 2$ belongs to **all** of the intervals $[y_1, y_2]$, since $\sqrt{1-\delta}+1 < 2 < \sqrt{1+\delta}+1$. If then the domain of ψ is enlarged to include $t = 0$, by taking $\psi(0) = 2$, the resulting function defined in this way is continuous at $t = 0$; in fact what we have just verified is that $\sqrt{1+t}+1$ is a continuous function — by filling in the missing value $\psi(0) = 2$ we have **made** $\psi(t)$ into a continuous function. In the notation of limits, what we have shown is that

$$\lim_{t \to 0} \frac{t}{\sqrt{1+t}-1} = 2$$

✸ *Example 9*

Consider two functions f_1, f_2, defined by $f_1(x) = \cos(1/x)$ and $f_2(x) = x\cos(1/x)$. In each case the domain is $\mathbb{R} \setminus \{0\}$, and we are interested in whether or not f_1 or f_2 converge to a limit as x tends to zero.

Consider first of all the function $\cos(1/x)$. For values of x in the interval $[-\delta, \delta]$, omitting of course $x = 0$, we know that $f_1(x)$ is in the range $-1 \leq f_1(x) \leq 1$. In fact, in this interval, there are lots of points x at which $f_1(x) = -1$, and lots of points at which $f_1(x) = +1$. (To make $f_1(x)$ equal to $+1$, take $x = 1/2N\pi$, so that $\cos(1/x) = \cos 2N\pi = 1$, where N is any positive integer which is large enough that $1/2N\pi < \delta$; to make $f_1(x)$ equal to -1, take $x = 1/[(2N+1)\pi]$.) Thus, for x in the interval $[-\delta, \delta]$, the shortest interval for which $y_1 \leq f_1(x) \leq y_2$ is $[y_1, y_2] = [-1, 1]$. Certainly the variation $V_0(\delta) = y_2 - y_1 = 2$ does **not** go to zero. Hence $\cos(1/x)$ does not have a limiting value as x tends to zero; the limit $\lim_{x \to 0} \cos(1/x)$ simply does not exist.

In the case of the second function $f_2(x) = x\cos(1/x)$, the situation is quite different. Here we know that $|f_2(x)| = |x\cos(1/x)| \leq |x|$, and in the interval $[-\delta, \delta]$ this leads to the inequality $-\delta \leq x\cos(1/x) \leq \delta$. The variation in $f_2(x)$, for x in this interval and $x \neq 0$, is 2δ, which **does** go to zero as δ is reduced to zero. It follows that $f_2(x)$ does have a limit as x tends to zero. Since $y_0 = 0$ is the only number for which the inequality $-\delta \leq y_0 \leq \delta$ holds for all $\delta > 0$, we see that the limit must be zero. In this case, then, we find $\lim_{x \to 0} x\cos(1/x) = 0$. By defining $f_2(0) = 0$, we can make f_2 into a continuous function with

$$f_2(x) = x \cos \frac{1}{x} \quad (x \neq 0)$$
$$= 0 \quad (x = 0).$$

EXERCISES ON 2.4

1. Choose some values of t close to zero at which to evaluate the function $\psi(t) = t/(\sqrt{1+t} - 1)$, using a calculator. Do your results appear to confirm that $\lim_{t \to 0} \psi(t) = 2$? Carry out a similar investigation of the function $\sin \theta / \theta$, for θ close to zero.

2. Using your calculator, investigate the behaviour of each of the functions $x \ln(|x|)$ and $|x|^x$ ($|x|$ raised to the power x), for x close to zero. What do your investigations suggest for the values of the limits $\lim_{x \to 0} x \ln(|x|)$ and $\lim_{x \to 0} |x|^x$?

3. Use problem 3 of Exercises on 2.3 to show that if $\lim_{x \to x_0} f(x) = l$ then $\lim_{x \to x_0} |f(x)| = |l|$. Think of an example to show that the converse of this result if **false**, in the sense that if $|g(x)|$ has a limit as $x \to x_0$, then it does not follow that $g(x)$ has a limit.

Summary

The concept of a function is basic to calculus, and we began by asking the question 'what is a function?' We came up with the answer that **any** rule for sending real numbers (input) to real numbers (output) defines a function, with the sole proviso that only a single output number is allowed for a given input number. For a function f, with input x and output y, we write this as $y = f(x)$ (but the rule could alternatively be written $y = g(t)$, or $z = \psi(s)$, or ...). The set of possible input numbers x is called the **domain** of f, and the set of possible output numbers y is called the **range** of f. Functions can often be expressed in terms of the successive application of a small number of basic operations, and we considered some examples of functions constructed in this way, and their graphs.

The notion of a function as I have described it is due to Dirichlet, who in 1837 proposed the definition of a function which is accepted today. In the early years of the 19th century, there had been no generally agreed idea of what mathematical functions were. The great German mathematician C.F. Gauss qualified his work on a class of function known as hypergeometric with the comment 'in as far as one can regard it as a function'. Subsequent progress during the century was often in the direction of exploring the properties of 'functions' which could behave more and more wildly – described as a 'gallery of monsters' and contrasted with the 'honest' functions which had preceded them. It was inevitable that these developments would lead to a new understanding of function as a much wider concept which embraced countless examples of pathology. Functions no longer need be described by single formulae, have smooth graphs, or even graphs that resemble curves at all. Functions need be ... just functions.

The ideas, techniques and applications of calculus which form the subject matter of this book will of necessity focus primarily on that core of functions which have tolerably smooth behaviour. Calculus may itself be fairly characterized, from a modern standpoint, as an attempt to bring a little order and regularity to the world of functions. Though content to remain for the most part within this well-ordered framework, we should also seek from time to time to glimpse the boundaries of our world, and to make ourselves aware of the wider, infinite universe beyond.

FURTHER EXERCISES

1. A function $f(x)$ is defined by $f(x) = \sqrt{9 - x^2}$.

   Evaluate $f(-2)$, $f(0)$ and $f(1)$. Sketch the graph of the function $f(x)$ and show that the graph is in the shape of a semicircle. What is the domain and range of this function?

2. Sketch the graph of each of the functions $1/(x^2 + 2)$, $x + |x|$, $x - \text{INT}(x)$, $x(x^2 - 1)$, $|x(x^2 - 1)|$ and $|x|(x^2 - 1)$.

3. A function y is said to be **even** if

   $t \in \text{domain}(y)$ implies $-t \in \text{domain}(y)$ and $y(-t) = y(t)$;

   y is said to be **odd** if

   $t \in \text{domain}(y)$ implies $-t \in \text{domain}(y)$ and $y(-t) = -y(t)$.

   Verify that even powers of t define even functions and odd powers of t define odd functions, but that some functions are neither even nor odd.

   Use the formula

   $$y(t) = \frac{y(t) + y(-t)}{2} + \frac{y(t) - y(-t)}{2}$$

   to show that any function y (for which $t \in \text{domain}(y)$ implies $-t \in \text{domain}(y)$) may be represented as the sum of an even and an odd function.

3 • Getting Functions Together

Many of the algebraic operations that you can carry out with numbers, such as multiply them, add, subtract, divide, you can also do with functions. If the function f is defined by the rule $f(x) = x^2$, and g is defined by $g(x) = \sin x$, then the sum $h = f + g$ of these two functions is defined in the obvious way by $h(x) = f(x) + g(x)$. For each input value, you simply add together the two output values, one for each function. Of course, this does not make sense unless x belongs to the domain of **both** functions. In general, the sum of two functions f and g will be given by $(f + g)(x) = f(x) + g(x)$, and the domain of $f + g$ will be the intersection of the respective domains of the functions f and g. In the same way, we can multiply, subtract or divide two functions. In the latter case, the domain of f/g will be slightly more complicated, because for x to be in the domain of f/g we have to stipulate not only that x is in the domain of f and the domain of g, but also that $g(x) \neq 0$.

In this chapter, we shall see how many of the functions with which we deal in calculus can be built up from quite simple functions which are then combined algebraically in various ways. Most functions can be evaluated from those available from a scientific calculator or computer, together with a small number of elementary operations. There is, in addition, one most important and fundamental way of combining functions, that of taking the composition. We shall find out what is meant by the composition of two functions, and how this notion can be used to define and evaluate inverse functions.

3.1 Ways of combining functions

To illustrate the ways in which the elementary algebraic operations may be applied to functions, it will be helpful to look again at some examples that we met in the last chapter. In Example 3 of Chapter 2, we sketched the graphs of the functions $|x + 1| + |x - 1|$ and $|x(1 - x)|$.

❋ Example 1

Consider first the function $|x + 1| + |x - 1|$. To evaluate this function, starting from a given input number, several basic operations are involved. In addition to the operation of taking the absolute value, represented by a box labelled ABS, we have to consider two operations, called, respectively, ADD1 and SUB1, or 'add one' and 'subtract one'. Apart from these, we shall have a box called SUM. The box SUM does exactly what it says — it sums or adds the numbers input into it. Hence SUM requires not one but two input numbers. Figure 3.1 represents the

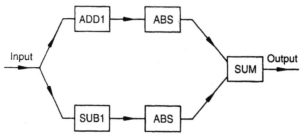

Fig 3.1   The function $|x + 1| + |x - 1|$.

function $|x + 1| + |x - 1|$ in terms of the various operations that we have defined. The input number follows two routes — the first route adds one and then takes the absolute value, the second subtracts one and takes the absolute value. The two numbers obtained in this way are then combined by SUM, and the resulting output is indeed their sum.

Most of the functions that we shall meet throughout this book may be conceived in this way, as resulting from a number of basic operations in combination. There are several advantages in representing functions like this. In the first place, by dissecting a function into its constituent components we shall more clearly understand how the function operates; the appearance of an elementary operation involving division, or the extracting of a square root for example, will draw attention to any restrictions on the domain of the function. The practical uses of 'flow diagrams' in the computer programming of function evaluation are well known, and it is often of importance to minimize or at least to control carefully the total number of basic operations involved. Finally, there are sound theoretical reasons for having recourse to representations of functions like that of Fig 3.1, as we shall have good reason to appreciate in Chapter 5 on differentiation.

✳ *Example 2*

As a second example from the last chapter, we may take the function $|x(1 - x)|$. Here we need a box called PROD, which takes two input numbers and evaluates their product. The flow diagram for this function is given in Fig 3.2.

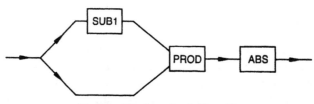

Fig 3.2   The function $|x(1 - x)|$.

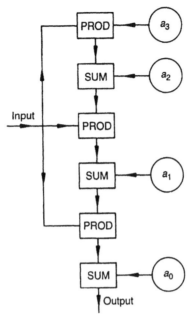

Fig 3.3 Function $a_3x^3 + a_2x^2 + a_1x + a_0$.

✳ *Example 3*

Cubic polynomial: for given coefficients a_3, a_2, a_1, a_0, define a polynomial of degree 3 by $p(x) = a_3x^3 + a_2x^2 + a_1x + a_0$.

Rather surprisingly, the evaluation of this function can be carried out by only six basic operations in all, each of which is either SUM or PROD, applied to a pair of numbers. To see how this is done, express the function, with liberal use of brackets, in the form

$$p(x) = x\{x(xa_3 + a_2) + a_1\} + a_0.$$

The evaluation of the function is sketched in Fig 3.3. The coefficients a_3, a_2, a_1, a_0 are fed in from the right. First multiply the input number by a_3; then add a_2; next multiply by the input number again and add a_1; finally, multiply by the input number and add a_0. The result of this sequence of six operations of multiplication and addition will be the required output number $p(x)$.

✳ *Example 4*

Let $g(t) = (\sqrt{t} + 1)/(\sqrt{t} - 1)$ be represented as in Fig 3.4, where SQRT denotes square root, and ADD1 and SUB1 are as before.

Here the box DIV requires two input numbers, which must be given in a specified order, and which are then divided. We adopt the convention that the number input at the upper arrow is divided **by** the number input at the lower arrow. If the lower input number is **zero**, then the division cannot be carried out, and a warning light comes on — we are outside the domain of the function g. This will happen if the number entering the SUB1 box is 1, i.e. if the input number,

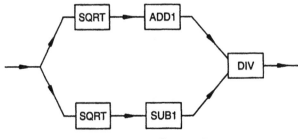

Fig 3.4 $g(t) = (\sqrt{t}+1)/(\sqrt{t} - 1)$.

which enters the SQRT box, is 1. The SQRT boxes will themselves malfunction if their input number is negative. All in all, then, the machinery which we have set up to calculate the function $g(t) = (\sqrt{t} + 1)/(\sqrt{t} - 1)$ will work perfectly well **unless** either $t < 0$ or $t = 1$. It follows that the domain of this function is the union of the two intervals $0 \leq t < 1$ and $1 < t < \infty$. We shall express this in set notation by writing the domain of g as $[0, 1) \cup (1, \infty)$.

EXERCISES ON 3.1

1. How many basic operations of addition and multiplication are needed to evaluate a general fourth degree polynomial $p(x) = \sum_0^4 a_n x^n$? Illustrate your answer by devising and implementing a strategy to evaluate a fourth degree polynomial of your own choosing. In the case $p(x) = 1 + 4x^2 + 4x^4$, show that only four operations are required.
2. Draw a 'flow diagram' to calculate the function $\sqrt{x}/(\sqrt{x - 1} - 1)$. Show how your diagram can be used to determine the domain of this function.
3. Choose some examples of functions that can be obtained from sequences of simple algebraic operations. Develop efficient strategies for evaluating these functions, and use flow diagrams to determine their respective domains.

3.2 Composition and inverse

The composition of two functions is nothing new. We have been doing this from the start of Chapter 2. There we looked at two functions, 'square' which took a number and squared it, and 'add two' which added two. If, as in Fig 2.1, we carry out these two operations one after another, then we have taken the **composition** of the two functions. Actually, as Fig 3.5 shows, the composition of 'square' and 'add

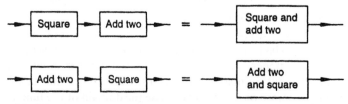

Fig 3.5 Composition of two functions — it depends on the order!

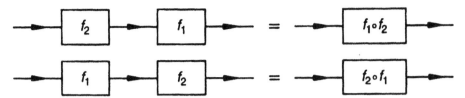

Fig 3.6 Composition of two functions f_1, f_2.

two' can be carried out in two different ways: 'square' and then 'add two' gives you the function $f(x) = x^2 + 2$; 'add two' and then 'square' gives you $f(x) = (x + 2)^2$. It will be useful to have a notation which distinguishes between these two different ways of taking the composition of a pair of functions.

Given functions f_1 and f_2, the composition that we get from letting f_2 operate first, then f_1, will be written as $f_1 \circ f_2$. If f_1 acts first, followed by f_2, this will be written as $f_2 \circ f_1$. Hence $f_1 \circ f_2$ is the function that sends x to $f_1(f_2(x))$, and $f_2 \circ f_1$ is the function that sends x to $f_2(f_1(x))$. This is often referred to as taking a 'function of a function'. The two compositions of f_1 and f_2 are then given by $f_1 \circ f_2$ and $f_2 \circ f_1$, where

$$(f_1 \circ f_2)(x) = f_1(f_2(x)); \quad (f_2 \circ f_1)(x) = f_2(f_1(x)).$$

Try to remember, when you see a composition $f_1 \circ f_2$, that it is the **second** function which operates **first**. The reason for this apparently illogical ordering is all due to the fact that our notation for functions demands that a function f stands to the left of the input value x. In following a historical convention that might well have turned out otherwise, we write $f(x)$, not $(x)f$. This means that in a queue of functions $f_1, f_2, f_3, \ldots$ waiting their turn in the evaluation of $f_1(f_2(f_3(x)))$ it is the functions further to the left which must wait the longer. The composition of three functions will be written as $f_1 \circ f_2 \circ f_3$, where $(f_1 \circ f_2 \circ f_3)(x) = f_1(f_2(f_3(x)))$, and the notation continues in an obvious way if yet more functions are introduced.

In taking the composition of a pair of functions, care must often be taken with domains. An input value x will belong to the domain of $f_1 \circ f_2$ whenever two conditions are satisfied. The first of these is that x belong to the domain of f_2; otherwise the input value will not get past the first hurdle — the evaluation of $f_2(x)$. Secondly, the value of $f_2(x)$ to be input into the function f_1 must belong to the domain of this function — provided $f_2(x)$ is in the domain of f_1 then the second hurdle has been passed. Thus

$$\text{domain}(f_1 \circ f_2) = \{x; x \in \text{domain}(f_2) \text{ and } f_2(x) \in \text{domain}(f_1)\}.$$

❋ *Example 5*

Let $f(x) = \cos x$ and $g(x) = x^2$. Then $(f \circ g)(x) = \cos(x^2)$, and $(g \circ f)(x) = \cos^2 x$. Clearly these functions are different. In general, $f \circ g \neq g \circ f$.

❋ *Example 6*

Let $f_1(t) = \sqrt{t}$, $f_2(t) = \sqrt{t-1}$. Then $f_1 \circ f_2 = u$, $f_2 \circ f_1 = v$, with $u(t) = (t-1)^{1/4}$ and $v(t) = \sqrt{t^{1/2} - 1}$. Both u and v have the domain interval $[1, \infty)$.

❋ *Example 7*

Let $f(x) = -\sqrt{x}$. Then $(f \circ f)(x) = -\sqrt{-x^{1/2}}$. The domain of f is the interval $[0, \infty)$ and the range of f is $(-\infty, 0]$. Only for $x = 0$ is $f(x)$ in the domain of f. Hence the domain of the composition $f \circ f$ of this function with itself consists of only the single point $x = 0$.

In defining the composition of two functions, I have simply given a name to a way of combining functions that is already familiar, and which arises naturally once we interpret functions as operations or rules for sending numbers to numbers. Another very closely related idea is that of taking the **inverse** of a function. Stated simply, if a function f sends 2 to 3, and -1 to π, and $\sqrt{2}$ to 1, ... , and a to $f(a)$, then its **inverse function** will send 3 to 2, and π to -1, and 1 to $\sqrt{2}$, ... , and $f(a)$ to a. The inverse function puts the operation of the function into reverse. It reverses the direction of the arrows, and undoes the effect of the function. Two examples of a function and its inverse are shown in Fig 3.7. The cube function sends any real number x into x^3. Hence the inverse of this function sends any real number x^3 into its cube root x.

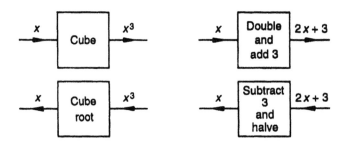

Fig 3.7   Two examples of a function and its inverse.

We shall often denote by f^{-1} the inverse of a function f. If $f(x) = x^3$ then $f^{-1}(x^3) = x$, so that $f^{-1}(x) = x^{1/3}$. The second example shown in Fig 3.7 is of the function $f(x) = 2x + 3$, for which $f^{-1}(x) = (x - 3)/2$.

Not every function has an inverse. The problem arises with any function for which two distinct values of x map into the same value of $f(x)$. Suppose, for two points x_1, x_2 in the domain of a function f, we have $x_1 \neq x_2$ but $f(x_1) = f(x_2) = c$, say. Then $f(x_1) = c$ would lead us to $f^{-1}(c) = x_1$, but $f(x_2) = c$ would give $f^{-1}(c) = x_2$. Since f^{-1} is supposed to be the inverse **function** of f, and since any function must send a number in its domain to a unique output number, in this case the inverse f^{-1} is not properly defined, and we have to say that such a function f has no inverse.

● *Definition I*

A function f is said to be **injective** if $f(x_1) = f(x_2)$ implies $x_1 = x_2$. The definition of an injective function means that no two distinct values of x give rise to the same

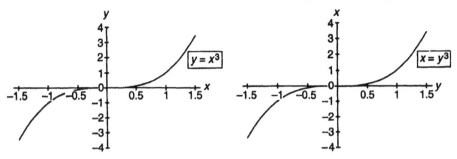

Fig 3.8 Graph of a function and its inverse.

value of $f(x)$. Equivalently, it means that any horizontal line will meet the graph $y = f(x)$ in at most one point. Every function f must satisfy the requirement that each value of x (in the domain of f) can give rise to just one value of y. An **injective** function satisfies the additional requirement that each value of y (in the range of f) can come from just one value of x.

The injective functions, also called one-to-one, are precisely those functions which have inverses. If f is an injective function, then its corresponding inverse function is defined by exchanging the roles of x and y. That is, if f is injective, then $y = f^{-1}(x)$ means the same as $x = f(y)$. The function f^{-1} is then obtained by solving for y in terms of x the equation $x = f(y)$. The function f has to be injective in order for the solution for y to be unique.

In Fig 3.8, we see the graph of the cubing function and its inverse. The cubing function is given by $y = x^3$. Its inverse is given by $x = y^3$, of which the solution for y is $y = x^{1/3}$. The graphs of the inverse function and of the original function are in fact identical, except for the labelling of the coordinate axes which has been changed to implement the reversal of roles $x \leftrightarrow y$. As a result of this relabelling, the coordinate axes in the second graph are not in their conventional positions. We can, of course, rotate this second graph in an anticlockwise direction through a right angle, so that the axis Oy points vertically upwards, but this brings the axis Ox to a leftward pointing direction, which is again counter to accepted conventions. Just one further rotation, after the first rotation through a right angle, will bring the axes back to their conventional positions. Hold the page, with the axis Oy pointing vertically upwards, and keeping Oy in this direction rotate the page about Oy through $180°$ until you can see the reverse side of the paper. (I have tried this experiment with a blackboard, but then it becomes more in the nature of a thought experiment. I have had more success with the use of an overhead projector.) Finally, hold the paper in this position up to the light. What you will see should look something like Fig 3.9, though I have taken the liberty of some minor changes in labelling.

As an alternative to the two rotations just carried out, it may be helpful to point out that their combined effect is that of a reflection of the original graph in the line $y = x$. This interpretation of the relation between the graphs of a function and of its inverse is illustrated in Fig 3.10, and holds quite generally for any injective function f.

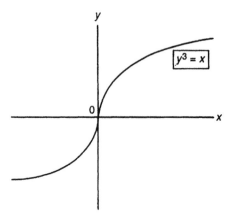

Fig 3.9 Graph of inverse function of x^3.

● *Example 8*

Let $f(x) = (x - 1)/(x + 1)$. The graph of f is given by $y = (x - 1)/(x + 1)$, and the graph of f^{-1} is $x = (y - 1)/(y + 1)$. Solving for y leads to $y = (x + 1)/(1 - x)$. So the inverse function is $f^{-1}(x) = (x + 1)/(1 - x)$. In this example, f has domain $\mathbb{R}\backslash\{-1\}$ and range $\mathbb{R}\backslash\{1\}$, whereas f^{-1} has domain $\mathbb{R}\backslash\{1\}$ and range $\mathbb{R}\backslash\{-1\}$.

● *Example 9*

Let $f(x) = x^2$. For the inverse function, we must solve for y the equation $x = y^2$. Since any positive value of x gives rise to **two** values of y, the inverse is not well

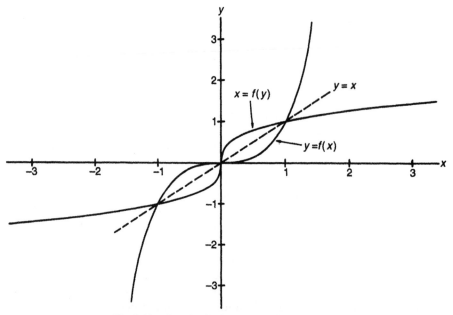

Fig 3.10 Graph of a function and its inverse.

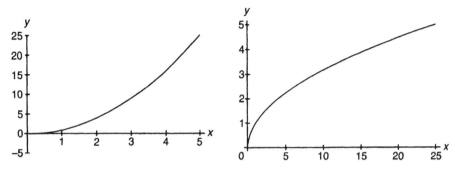

Fig 3.11 Graph of x^2, restricted to $[0, \infty)$ ... and of its inverse function.

defined. In fact, the function $f(x) = x^2$ is not injective, since we can have $f(x_1) = f(x_2)$ without $x_1 = x_2$. Rather than giving up completely on an inverse, the next best thing is to find an interval of x values on which $f(x) = x^2$ **is** injective. If, then, we narrow down the domain of the function to consist of just this interval, then we have an injective function and an inverse can be defined without difficulty. A quick look at the graph of the function x^2 should convince you that the best (i.e. largest) intervals for this purpose are either $(-\infty, 0]$ or $[0, \infty)$.

Consider, for example, the function $f(x) = x^2$ restricted to the interval $[0, \infty)$. The graph of this function is shown in Fig 3.11, and is given by $y = x^2$ $(x \geq 0)$. Note that, with the domain restricted to the positive real line, a horizontal line will meet the graph in at most one point, as required in the case of an injective function. Exchanging x and y, the graph of the inverse function is now given by $x = y^2$ $(y \geq 0)$, of which the solution for y is $y = +\sqrt{x}$. I have written $+\sqrt{x}$ to emphasize the sign, but $\sqrt{x}$ means the positive square root in any case, so $y = \sqrt{x}$ will do just as well; so by restricting the domain of x^2 to $[0, \infty)$, the inverse function becomes just the square root function. This is of course very much to be expected — to invert the square of a **positive** number, just take the (positive) square root.

Had we made the alternative choice of restricting the domain of x^2 to $(-\infty, 0]$, then again we should have an injective function, the inverse in this case being the function $-\sqrt{x}$.

❧ *Example 10*

I have tried to stress the need for a function to be injective or one-to-one, in order to have a proper inverse, so it is a little disconcerting to realize that many of the standard inverse functions offered by the scientific calculator or mathematical computer packages relate to non-injective functions. The trigonometric functions sin x, cos x and tan x are examples of periodic functions. A function f is said to be **periodic** if there is a positive number L such that $f(x + L) = f(x)$ for all x in the domain of f; the smallest positive value of L is then called the **period** of f. This means that sin x and cos x have period 2π, and tan x has period π. [In principle a function may be periodic but with **no** smallest possible value of L. This would happen, for example, if $f(x + L) = f(x)$ for $L = 1/2^N$ for all positive integer values of N; in that case L can be as small as you like but has no smallest **positive** value.

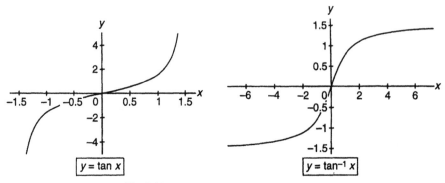

Fig 3.12 Tan function and its inverse.

Functions having this property are either discontinuous at all points of their domain, or constant functions, and we shall not consider them further here.]

A periodic function cannot be injective, since $x_2 = x_1 + L$ will allow $f(x_1) = f(x_2)$ but $x_1 \neq x_2$. However, just as in Example 9, we can look for intervals on which f is injective. Consider first of all the tan function, which is injective when restricted to the interval $-\pi/2 < x < \pi/2$. Figure 3.12 shows the graph $y = \tan x$ $(-\pi/2 < x < \pi/2)$ as well as the graph of its inverse function $x = \tan y$ $(-\pi/2 < y < \pi/2)$.

Denoting by f^{-1} the inverse of a function f, we shall write $\tan^{-1}$ for the inverse tan function. The domain of $\tan^{-1}$ is $\mathbb{R}$, and the range is the interval $(-\pi/2, \pi/2)$. Define $\tan^{-1}$ to be the unique value of y in the interval $-\pi/2 < y < \pi/2$ for which $\tan y = x$.

The inverse sine and cosine function can be treated in a similar way. The sine function is injective on the interval $-\pi/2 \leq x \leq \pi/2$, and the cosine function on the interval $0 \leq x \leq \pi$. This allows us to define corresponding inverse functions $\sin^{-1}$ and $\cos^{-1}$. To summarize, we have

$$y = \tan^{-1} x \Leftrightarrow x = \tan y \quad (-\frac{\pi}{2} < y < \frac{\pi}{2})$$

$$y = \sin^{-1} x \Leftrightarrow x = \sin y \quad (-\frac{\pi}{2} \leq y \leq \frac{\pi}{2})$$

$$y = \cos^{-1} x \Leftrightarrow x = \cos y \quad (0 \leq y \leq \pi)$$

Alternative names exist for each of these three inverse trigonometric functions; they are called, respectively, arctan, arcsin and arccos. Notice that only the first of the three, the inverse tangent function, is defined for all $x \in \mathbb{R}$. For the inverse sine and cosine function, the domain in each case is the interval $[-1,1]$.

⊕ *Example 11*

Assuming the identity $\cos 2\theta = 2\cos^2 \theta - 1$, deduce a corresponding identity for the function $\cos^{-1}(\sqrt{x/2})$. For which set of x values does this identity hold?

This example illustrates how it is often possible to convert properties of functions into corresponding results for their inverse functions. Note first of all that $\cos^{-1}(\sqrt{x/2})$ makes sense only if $\sqrt{x/2}$ belongs to the interval $[-1, 1]$. Since $\sqrt{x/2}$ is non-negative, this requires in fact that $\sqrt{x/2}$ belongs to $[0,1]$, so that the domain of the function is the interval $0 \le x \le 2$.

Now $y = \cos^{-1}(\sqrt{x/2})$ is equivalent to $\cos y = \sqrt{x/2}$. For the proper definition of the inverse cosine, we need $0 \le y \le \pi$. However, here we require $\cos y$ to lie between 0 and 1, so only $0 \le y \le \pi/2$ is allowed. 'Solving' for x, we have $x = 2\cos^2 y$, which from the stated identity leads to $x = 1 + \cos 2y$ $(0 \le y \le \pi/2)$. Hence $\cos 2y = (x - 1)$ $(0 \le 2y \le \pi)$, which again from the definition of inverse cosine allows us to conclude that $y = \frac{1}{2}\cos^{-1}(x - 1)$.

Hence the required identity is $\cos^{-1}(\sqrt{x/2}) = \frac{1}{2}\cos^{-1}(x - 1)$, and holds for all x in the interval $0 \le x \le 2$.

Example 12

Let $f(x) = x \sin x$ $(0 \le x \le \pi/2)$. As x increases from 0 to $\pi/2$, the value of $f(x)$ increases from 0 to $\pi/2$. This function is injective on the given interval, and the inverse function $y = f^{-1}(x)$ is to be obtained by solving for y the equation

$$x = y \sin y \quad (0 \le y \le \pi/2)$$

The solution of this equation for y can only be carried out numerically, and there is no simple expression for y in terms of standard functions. In such a case, we must be content with merely writing f^{-1} for the inverse function. It should, after all, be realized that none of the functions $\sin^{-1}$, $\cos^{-1}$, $\tan^{-1}$ (or sin, cos, tan for that matter) can be calculated exactly, except for certain special values of x. These are merely convenient notations for a class of functions which it will prove extremely useful to have added to our vocabulary, and about which we will learn a great deal more in due course.

How does the inverse of a function relate to some of the other operations with functions, which we looked at earlier in this chapter? In fact the inverse has very little to do with the various algebraic operations such as addition and multiplication of functions. There is no simple way, for example, of writing down the inverse of the sum or product of two functions, in terms of the individual inverses. Certainly the very greatest care must be taken to distinguish between f^{-1} (inverse function of f, as defined in this section) and $1/f$ (the algebraic operation $1 \div f$). If I wish to refer to $1 \div f(x)$, then I can write $1/f(x)$, or even $(f(x))^{-1}$, but **never** $f^{-1}(x)$. In particular, $\tan^{-1}(1) = \pi/4$, but $(\tan(1))^{-1}$ is $1/\tan(1)$, or $1 \div \tan(1)$, which is of course quite different.

Where there is a useful interplay between inverse functions and other operations with functions concerns the **composition** of two or more functions. Noting that the definition of inverse functions can be summarized in the equations

$$f^{-1}(f(x)) = x \quad (x \in \text{domain}(f))$$

and

$$f(f^{-1}(x)) = x \quad (x \in \text{domain}(f^{-1}))$$

and defining an identity function I by $I(x) = x$ (having the same domain as f in the first instance, and the same domain as f^{-1} in the second instance) we can use the notation for composition to great effect by writing

$$f^{-1} \circ f = I$$

and

$$f \circ f^{-1} = I$$

For the inverse of the composition of two functions f and g, we have the useful rule $(f \circ g)^{-1} = g^{-1} \circ f^{-1}$. To overcome any surprise that may be felt on seeing $g^{-1} \circ f^{-1}$ rather than $f^{-1} \circ g^{-1}$ on the right-hand side, it may be worth noting that you cannot **take off** your shoes and socks in the same order that you put them on!

EXERCISES ON 3.2

1. By inventing examples of your own, convince yourself that it is **not** true in general that $f \circ g = g \circ f$, for the composition of two functions f and g. Make up some examples of pairs of functions for which it **is** true that $f \circ g = g \circ f$. (For example, look at powers, and **some** linear functions.)
2. Functions f and g are defined by $f(x) = x^2$, $g(x) = \cos x$. Use the notation for compositions to express each of the functions $(\cos x^2)^2$, $\cos(x^4)$ and $(\cos x)^4$ in terms of compositions of f with g.
3. A function g is defined by $g(t) = (1 - t)/(1 + t)$. Show that $g = g^{-1}$, i.e. this function is its own inverse. Try to find other functions which equal their own inverse functions.
4. Use a flow diagram to explain why $(f \circ g)^{-1} = g^{-1} \circ f^{-1}$.
5. Paying careful attention to signs, show that $\cos(\sin^{-1} x) = \sin(\cos^{-1} x) = \sqrt{1 - x^2}$ $(-1 \le x \le 1)$. Show that $\cos(\tan^{-1} x) = 1/(\sqrt{1 + x^2})$, for all $x \in \mathbb{R}$.

Summary

Starting from a small collection of basic functions, such as powers, trigonometric functions, the modulus function, later to be supplemented by the exponential, logarithm, and related functions, we can combine them in various ways to make new functions, and thereby very considerably enlarge our collection. For this reason, anyone with a scientific or programmable calculator, or computer, will have access numerically to a huge variety of different functions. Algebraically, functions can be combined by adding them, subtracting them, multiplying, or dividing. We can also take the composition of two functions, which means letting one of the functions operate, and then the other. Related to this is the idea of taking the **inverse** of a function. Not all functions have an inverse, and those which do are called injective (or one-to-one). To be injective, a function must not assume the same value, $f(x)$, for two different values of x.

Functions which are built up from simpler functions in this way are often well described by means of flow diagrams. We have seen several examples of flow diagrams so far. A flow diagram may be a very useful aid to computing the values of a function, and is often the starting point for writing a computer program to carry out this evaluation. In addition, a flow diagram will draw attention to any domain restrictions, and will clearly exhibit where the evaluation of one function may depend on another. Later we shall see the useful role of flow diagrams in clarifying the process of differentiation for more functions in combination.

FURTHER EXERCISES

1. For which values of x is it true that $(1+x)/(1+2x) \geq 0$? Draw a flow diagram for the function $\sqrt{(1+x)/(1+2x)}$. What are the domain and range of this function?

2. Using only the basic operations 'square' and 'add one', show how to evaluate the function $x^2 + 2x + 4$. If $f =$ 'square' and $g =$ 'add one', use the notation for composition of functions to express the function $x^2 + 2x + 4$ in terms of compositions of f with g.

3. If $f_1(x) = \sqrt{x}$ and $f_2(x) = x^2$, what function is the composition $f_1 \circ f_2$?

4. Find the inverses of the functions $1/x$, $(x^3 + 1)/(x^3 - 1)$ and $2\sqrt{x} + 1$, stating in each case the domain and range of both function and inverse.

5. Prove the identity $\sin^{-1} x + \cos^{-1} x = \pi/2$. The secant function is defined by $\sec x = 1/\cos x$. Define a function $\sec^{-1} x$, and show that $\sec^{-1} x = \cos^{-1}(1/x)$.

6. (a) A function f is given by $f(t) = 1 + (1/t)$.
   Write down expressions for $(f \circ f)(t)$, $(f \circ f \circ f)(t)$ and $(f \circ f \circ f \circ f)(t)$, and explain how the pattern continues.
   (b) A sequence $x_1, x_2, x_3, \ldots$ is defined by $x_1 = 1$, $x_{n+1} = 1 + (1/x_n)$. Evaluate $x_2, x_3, x_4, \ldots$. Assuming that the sequence converges to a limit x, use a calculator to estimate this limit. Show that the limit satisfies the equation $x^2 - x - 1 = 0$, and hence determine x. Explain the connection between (a) and (b) of this exercise.

4 • Rates of Change, Slopes, Tangents

Of all functions, the simplest are the **constant** functions. In fact, for the very reason that they **are** so simple, constant functions are often overlooked. A constant function is a function which sends every number in its domain into the same output number. If we denote by the letter b this common output number, and the function itself by f, then our constant function will be given by the rule $f(x) = b$, and its graph will be the horizontal straight line $y = b$. Constant functions have the property that any change in input value will not give rise to any corresponding change in output value.

After the constant functions come the **linear** functions. A linear function is a function f defined by a rule of the form $f(x) = ax + b$, where a and b are given real numbers. The graph of such a function is a straight line $y = ax + b$ having slope or gradient a, and with intercept $y = b$ on the y-axis. Any straight line, except for a vertical straight line, is the graph of a linear function, and the special case $a = 0$ for which the slope is zero reduces again to a constant function.

The characteristic property of linear functions is that any change in input number will give rise to a corresponding and **proportionate** change in output number. For example, if $y = ax + b$, a change in input value from x_1 to x_2 will produce a change in output value from y_1 to y_2, where

$$(y_2 - y_1) = (ax_2 + b) - (ax_1 + b) = a(x_2 - x_1)$$

So, for a linear function, the slope or gradient of the graph may also be thought of as a **rate of change**, allowing a comparison between change in output and change in input, and given by

$$a = \frac{y_2 - y_1}{x_2 - x_1} = \frac{\text{change in output}}{\text{change in input}}$$

As a straightforward example of this relationship connecting outputs and inputs for linear functions is the rule relating degrees Celsius t_C to degrees Fahrenheit t_F, given by $t_C = 5(t_F - 32)/9$. Given a change in temperature on the Fahrenheit scale, multiplication by the factor $5/9$ will produce a corresponding change on the Celsius scale.

In this chapter, we shall begin to see how to extend the idea of rates of change of functions, and the closely related idea of gradient of a graph, to functions which will not necessarily be linear. Our main concern will be with the class of functions, known as **differentiable**, for which the proportionality between output change and input change no longer continues to hold for all possible inputs and all possible outputs, but in a limiting sense as a **local approximation**. Looked at another way, the graph of a differentiable function will be locally linear, in the sense that it will look more and more like a straight line as we approach shorter and shorter arcs of the graph. Locally, then, we shall find that differentiable functions behave like linear functions, and their graphs approximate to straight lines. Mathematics, as so often happens, here too yields a useful economy of ideas, in that the three subjects of this chapter, rates of change, slopes and tangents, will be brought together under the single heading of differentiation.

4.1 Rates of change and gradients

❧ *Example 1*

To illustrate some of the ideas of this chapter, we consider the function f defined by $f(x) = x^2$.

A change in input value from, say, x_1 to x_2 will give rise to a corresponding change in output value from y_1 to y_2, where

$$(y_2 - y_1) = f(x_2) - f(x_1) = x_2^2 - x_1^2 = (x_2 + x_1)(x_2 - x_1)$$

Unlike the case of a linear function, a comparison between change in output and change in input here yields $(y_2 - y_1)/(x_2 - x_1) = x_2 + x_1$, which is **not** a fixed ratio but depends on the respective values x_1, x_2 of the input numbers.

Locally, however, we are often interested in **small** changes in input from x_1 to x_2, so that x_1 will be close to x_2, and x_1 and x_2 both close to some fixed number x in the domain of f. We are then concerned with the variation in the function in the neighbourhood of x, and in that case, for x_1 and x_2 close to x, the ratio $(y_2 - y_1)/(x_2 - x_1)$ will closely approximate to $2x$.

Looked at from a more geometrical point of view, suppose, in Fig 4.1, we have to determine the value of the slope or gradient of the graph $y = x^2$ at some given point P. A natural first step is to take points X_1, X_2 of the graph, with X_1 to the left of P, and X_2 to the right. As the points X_1, X_2 are taken closer to P, we would expect the gradient of the straight line X_1X_2 to approach the gradient of the graph at the point P. Indeed, this limiting gradient of X_1X_2, as X_1 and X_2 approach P, can be taken to be **what we mean by** the gradient of the graph at P.

If X_1 has Cartesian coordinates (x_1, y_1), and X_2 has Cartesian coordinates (x_2, y_2), so that $y_1 = x_1^2$ and $y_2 = x_2^2$, the gradient of the straight line X_1X_2 will then be

$$\frac{y_2 - y_1}{x_2 - x_1} = \frac{x_2^2 - x_1^2}{x_2 - x_1} = x_2 + x_1$$

which approaches $2x$ in the limit as x_2 and x_1 approach the x coordinate of the point P. So we can say that the gradient at a general point (x, x^2) of the graph is just $2x$.

More generally, the gradient at a point (x, y) of a graph $y = f(x)$ may be defined as the limiting value of the ratio $(y_2 - y_1)/(x_2 - x_1)$ as x_2 and x_1 both approach x with $x_1 \leq x$, and $x_2 \geq x$, and $y_1 = f(x_1), y_2 = f(x_2)$. To determine this gradient we have to evaluate the ratio $[f(x_2) - f(x_1)]/(x_2 - x_1)$ as x_1 and x_2 approach x, x_1 from the left and x_1 from the right. We allow as special cases $x_1 = x$ (corresponding to X_1 at P in Fig 4.1), or $x_2 = x$ (corresponding to X_2 at P). **Not**, of course **both** $x_1 = x$ and $x_2 = x$, which would make the ratio $(y_2 - y_1)/(x_2 - x_1)$ meaningless.

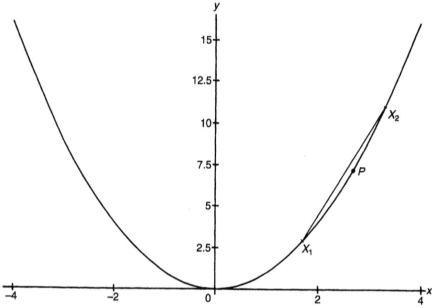

Fig 4.1   Geometrical interpretation of gradient.

Indeed, there is advantage in paying special attention to the cases $x_1 = x$ and $x_2 = x$, respectively. This is because if each of the ratios $r_1 = [f(x) - f(x_1)]/(x - x_1)$ and $r_2 = [f(x_2) - f(x)]/(x_2 - x)$ approach a common limit l, respectively, as x_1 tends to x $(x_1 < x)$ and x_2 tends to x $(x_2 > x)$, then the ratio $r = [f(x_2) - f(x_1)]/(x_2 - x_1)$, which may be written as $[(x - x_1)r_1 + (x_2 - x)r_2]/(x_2 - x_1)$, will also approach the limit $[(x - x_1)l + (x_2 - x)l]/(x_2 - x_1) = l$.

A common notation is to set $x_1 = x + h$ or $x_2 = x + h$, and to treat the two cases on a common footing with $h < 0$ in the case of x_1 and $h > 0$ for x_2. The ratios r_1 and r_2 are then each of the form $[f(x + h) - f(x)]/h$. The gradient of $y = f(x)$ is then the limiting value of this ratio as h approaches zero. In this notation, h tending to zero through positive values corresponds to the previous limit as x_2 tends to x, and h tending to zero through negative values corresponds to the previous limit as x_1 tends to x. These two limits, respectively with h positive and h negative, must be equal. We have come here to the formal definition of the gradient, or derivative, of a function. Though often convenient in practice, and certainly an important starting point for a great deal of rigorous analysis, this definition is just a reformulation of the original idea of rate of change or gradient, from which we started. Since my intention, rather than to make inroads into formal analysis (itself the subject of a separate volume in this series), is to convey the essential ideas of the subject and the use of these ideas in applications, we leave to other texts the deeper treatment of notions such as limit, continuity, and so on, which underlie the concept of derivative. It is to be hoped that the committed reader will, in due course, return to these questions, which from a formal point of view underpin the subject.

● *Definition I*

The **right derivative** of a function f, at a point x in the domain of f, is the limiting value of the ratio $[f(x+h) - f(x)]/h$, as h approaches zero through positive values. This limit, whenever it exists, will be denoted by $\lim_{h \to 0+}[f(x+h) - f(x)]/h$. The **left derivative** of f, at x, is the limit of the same ratio, as h approaches zero through negative values, and will be denoted by $\lim_{h \to 0-}[f(x+h) - f(x)]/h$. If both left and right derivatives exist at x, and are equal, then f is said to be **differentiable** at x. The common limit, in that case, is called the **derivative of f, at** x, and will be written as $f'(x)$, where

$$f'(x) = \lim_{h \to 0} \frac{f(x+h) - f(x)}{h} \tag{4.1}$$

A function f which is differentiable at each point of its domain will be called a **differentiable** function, and the function f' defined by equation (4.1) will be called the **derivative** of f.

The right and left derivatives of a function at a point are often used in cases where the domain of the function f is an interval, or union of intervals. For example, if $f(x) = x^2$ with domain the interval $-1 \le x \le 1$, then f is differentiable at each x in the open interval $-1 < x < 1$, with derivative $f'(x) = 2x$, whereas at $x = -1$ the right derivative is -2 and at $x = 1$ the left derivative is $+2$.

Following our previous discussion, the derivative of f at x may be identified with the **rate of change** of f, at x. Alternatively, $f'(x)$ is the gradient of the graph $y = f(x)$ at the point $(x, f(x))$. We shall have more to say with regard to each of these interpretations of the derivative.

⊞ *Example 2*

Define a function u, with domain $\mathbb{R} \setminus \{0\}$, by

$$u(t) = t^3 + 1/t$$

Equation (4.1), adapted to this notation, becomes

$$u'(t) = \lim_{h \to 0} \frac{u(t+h) - u(t)}{h}$$

where

$$u(t+h) = (t+h)^3 + \frac{1}{t+h}$$

$$= t^3 + 3t^2 h + 3th^2 + h^3 + \frac{1}{t+h}$$

so that

$$\frac{u(t+h) - u(t)}{h} = \frac{t^3 + 3t^2 h + 3th^2 + h^3 + 1/(t+h) - t^3 - 1/t}{h}$$

$$= \frac{3t^2 h + 3th^2 + h^3 - h/t(t+h)}{h}$$

$$= 3t^2 + 3th + h^2 - \frac{1}{t(t+h)}$$

which, in the limit as $h \to 0$ (through positive or negative values), approaches $3t^2 - 1/t^2$ provided $t \neq 0$.

Since the derivative of u exists at all t in the domain of u (that is, at all $t \neq 0$), u is a differentiable function, with derivative given by

$$u'(t) = 3t^2 - 1/t^2$$

The same result will be obtained if we evaluate the limit using a pair of values t_1, t_2 in the domain of u, with $t_1 \leq t \leq t_2$ and let t_1 and t_2 tend to t. In that case, with $u_2 = u(t_2)$ and $u_1 = u(t_1)$, we have

$$\frac{u_2 - u_1}{t_2 - t_1} = \frac{(t_2^3 + 1/t_2) - (t_1^3 + 1/t_1)}{(t_2 - t_1)}$$

Using the identity $(t_2^3 - t_1^3) = (t_2 - t_1)(t_2^2 + t_1 t_2 + t_1^2)$, the ratio simplifies to $t_2^2 + t_1 t_2 + t_1^2 - 1/t_1 t_2$, which again gives the limit $3t^2 - 1/t^2$ as t_1, t_2 approach t, for $t \neq 0$.

✦ Example 3

The function $f(x) = \sin x$ is known to satisfy the trigonometric identity

$$\sin A - \sin B = 2 \sin\left(\frac{A - B}{2}\right) \cos\left(\frac{A + B}{2}\right)$$

It is also known that $\sin \theta / \theta$ tends to unity as θ tends to zero. Here θ must be in radians. We can use these two properties of the sine function to determine its derivative.

Thus

$$\frac{f(x_2) - f(x_1)}{(x_2 - x_1)} = \frac{\sin x_2 - \sin x_1}{x_2 - x_1}$$

$$= 2 \sin\left(\frac{x_2 - x_1}{2}\right) \cos\left(\frac{x_2 + x_1}{2}\right) \Big/ (x_2 - x_1)$$

$$= \left\{ \sin\left(\frac{x_2 - x_1}{2}\right) \Big/ \left(\frac{x_2 - x_1}{2}\right) \right\} \cos\left(\frac{x_1 + x_2}{2}\right).$$

The expression in the braces is just $\sin \theta / \theta$, where $\theta = (x_2 - x_1)/2$ approaches zero in the limit as x_1, x_2 tend to x; hence this expression tends to unity in the limit. In the limit $x_1 \to x$, $x_2 \to x$, we also have $\cos((x_1 + x_2)/2) \to \cos x$. Hence $f(x) = \sin x$ is a differentiable function, and we have in this case $f'(x) = \cos x$. The same conclusion follows, of course, if we use equation (4.1) to express the derivative as a limit $h \to 0$.

✦ Example 4

The following example shows that not all functions are differentiable.

Let f be the modulus function $|x|$. To evaluate the derivative of f, we have

$$f'(x) = \lim_{h \to 0} \frac{|x + h| - |x|}{h}$$

Since for $x > 0$ we have $x + h > 0$ for $|h|$ small enough (in particular for $|h| < x$), for positive x we can replace $|x + h|$ by $x + h$ in the limit, and $|x|$ by x, so that

$$f'(x) = 1 \quad \text{for} \quad x > 0$$

Similarly,

$$f'(x) = -1 \quad \text{for} \quad x < 0$$

However, for $x = 0$ we should have

$$f'(0) = \lim_{h \to 0} |h|/h$$

and this limit does not exist. The right derivative $\lim_{h \to 0+} |h|/h$ exists and is 1; the left derivative also exists and is -1. Since these two limits exist but are different, the function $|x|$ is not differentiable at $x = 0$. We say that $|x|$ is a non-differentiable function. This conclusion is not unreasonable if we consider the graph of the function near $x = 0$. Another aspect of this non-differentiability of the modulus function is that the ratio $(|x_2| - |x_1|)/(x_2 - x_1)$, with $x_1 \leq 0 \leq x_2$, in the limit $x_1 \to x, x_2 \to x$, may be made to approach any assigned real number in the interval $[-1,1]$, by suitable choice of the pair x_1, x_2. For example, taking $x_1 < 0 < x_2$ and letting x_1, x_2 tend to zero with $x_2 = -x_1$ the limit is zero, whereas with $x_2 = -2x_1$ the limit is $1/3$.

The rate of change of the modulus function is therefore undefined at $x = 0$, though at all other points one has $f'(x) = x/|x|$.

EXERCISES ON 4.1

1. A malevolent demon attaches a giant cycle pump to the earth, and pumps the earth up until its radius increases by 1 m. Treating the earth as a perfect sphere having a diameter of about 12,700 km, determine the resulting increase in the earth's (a) circumference, (b) surface area, and (c) volume, and explain in each case the role of rate of change in your calculation.

2. If $f(x) = \sin x$ (with x in radians), use your calculator to evaluate $[f(x_2) - f(x_1)]/(x_2 - x_1)$ for various pairs x_1, x_2 with x_1 close to x_2, say $|x_2 - x_1|$ of the order $10^{-1}, 10^{-2}, 10^{-3}$, etc. Verify that your results are consistent with $f'(x) = \cos x$.

   Carry out a similar investigation for the function $u(t) = t^3 + 1/t$.

3. Explain why any **even** differentiable function has $f'(0) = 0$. If f is an even function, what is the relation between $f'(x)$ and $f'(-x)$? How are $f'(x)$ and $f'(-x)$ related if f is an **odd** function? (For the definition of even and odd functions, see exercise 3 at the end of Chapter 2.)

4. By considering carefully the separate cases $x > 0$, $x = 0$ and $x < 0$ and using equation (4.1), show that the function $f(x) = x|x|$ is differentiable with $f'(x) = 2|x|$. Use **the respective graphs** of the functions $x|x|$ and $|x|$ to explain why the first of these two functions is differentiable but the second is not.

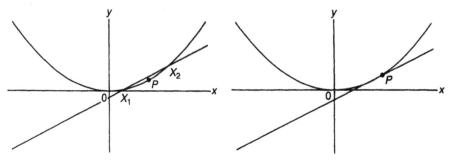

Fig 4.2 Tangent of a graph at a point.

4.2 Tangents and linear approximation

We have seen, with reference to the graph $y = x^2$ of the function $f(x) = x^2$, how the gradient of a chord $X_1 X_2$ approaches a limit as X_1, X_2 tend to a given point P of the graph. This limit, by definition, is the gradient of the graph at P. The chord $X_1 X_2$ itself, extended as in Fig 4.2, approaches the tangent at P to the graph.

Let us determine the equation of the tangent at a general point (x_0, y_0) of the graph. At the point $P = (x_0, y_0)$, the gradient is $f'(x_0) = 2x_0$. Evidently then the tangent, having slope $2x_0$, must have an equation of the form

$$y = 2x_0 x + b$$

However, the tangent passes through the point (x_0, y_0). Hence $y_0 = 2x_0^2 + b$, giving $b = y_0 - 2x_0^2$. Thus we have, for the equation of the tangent at (x_0, y_0) in this case:

$$y - y_0 = 2x_0(x - x_0)$$

In terms of x_0 alone, substituting $y_0 = x_0^2$ gives for the tangent $y = 2x_0 x - x_0^2$.

Similar arguments apply in the case of any differentiable function $f(x)$, and the tangent at a general point (x_0, y_0) of the graph $y = f(x)$ has the equation

$$y - y_0 = f'(x_0)(x - x_0) \tag{4.2}$$

The tangent to a curve at a point P may be regarded as the best linear approximation to the curve in the neighbourhood of P. That is, the tangent is the closest approximation of the curve near P by a straight line. Put another way, we can say that locally, differentiable functions look like linear functions.

This property of differentiable functions is illustrated in Fig 4.3 for the function $f(x) = x(x + 2)$. To investigate the behaviour of this function near $x = 0$, the graph is successively magnified about the origin by a scaling factor M.

Two successive stages in this process are shown, for increasing values of the scaling factor. The graph on the right, for a suitably large value of M, shows that under such high magnification the part of the curve very close to the origin is barely distinguishable from a straight line. The centre graph shows an intermediate stage. As M increases and we move to the right in the sequence of graphs under successively higher magnification, the arc of the curve depicted corresponds to an ever shorter arc of the original graph for the function $x(x + 2)$.

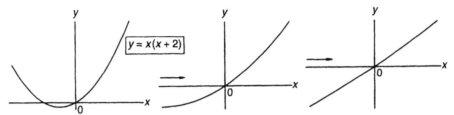

Fig 4.3   Effect of scaling on a differentiable function.

The use of change of scale to investigate the local behaviour of functions can also be carried out analytically. Consider, for example, the graph $y = f(x)$ of a differentiable function f for which $f(0) = 0$. (This graph passes through the origin, but by suitable choice of origin we can reduce the more general situation to this special case.) Now magnify the graph by a scale factor 2 about the origin. This means that any point (x_0, y_0) of the original graph, for which $y_0 = f(x_0)$, becomes a point $(2x_0, 2y_0)$ of the magnified graph. Since $2y_0 = 2f(2x_0/2)$, the equation of the magnified graph must be $y = 2f(x/2)$, and in the same way, an enlargement by scale factor M gives rise to a graph $y = Mf(x/M)$. In the example illustrated by Fig 4.3, for which $f(x) = x(x + 2)$, corresponding to a scale factor M we find $y = M(x/M)[(x/M) + 2] = x[2 + (x/M)]$, which, as we expect, approaches for large M the linear function $y = 2x$. Here the gradient 2 of the limiting straight line is identical to the gradient of the graph at the origin of the original function; that is $2 = f'(0)$. We can understand why this is so by using equation (4.1) to evaluate the derivative at the origin, giving $f'(0) = \lim_{h \to 0} f(h)/h$, which on substituting $h = x/M$ and proceeding to the limit $M \to \infty$ results in $\lim_{M \to \infty} (M/x)f(x/M) = f'(0)$. Hence the scaled function $Mf(x/M)$ does indeed asymptotically approach the linear function $f'(0)x$.

The behaviour of differentiable functions under scaling is in complete contrast to that of non-differentiable functions. For example, it is easy to see that a scaling transformation by an arbitrary factor M about the origin leaves the non-differentiable function $f(x) = |x|$ invariant. That is, the scaled function $Mf(x/M)$ is identical to the original function, and no degree of scaling will remove the sharp kink in the graph at $x = 0$. The modern theory of fractals is rich with examples of functions which are not differentiable at **any** point of their graph. Indeed, a basic principle governing the construction of many of these examples is to consider their transformation under scaling. An important idea is that of self-similarity — a fractal curve under suitable scaling, often coupled with translation or rotation or both, may map on to some part of itself. This gives rise to curves which, far from becoming simpler under magnification, as is the case for differentiable functions, may exhibit a higher level of complexity as successive scales of magnification are reached.

Let us return, then, to the more regular and smoother world of differentiable functions, and note that their approximate linearity in the local domain makes such functions of special importance in the description of a wide variety of physical processes. It is not our main intention in this book to outline the role of differentiation in the formulation of scientific theories. Certainly the development of calculus and of physical science have been inextricably linked since the time of Newton. Every major scientific advance, from Maxwell's electromagnetic theories of the 19th century to Einstein's establishment of the principles of special and general relativity in more modern times, has used the language of differential equations to express its deeper truths.

At a more mundane level, one may expect calculus to be central to any attempt to relate one physical quantity to another by physical law. It is fundamental to scientific method itself, for example in observing the extension of a stretched spring as given weights are attached to it, that the experimenter controls the parameters of the experiment by looking for a relationship between extension and attached weight over a **small range** of these quantities. This leads naturally to the modelling of physics through differentiable functions; for example, the approximate linearity relating extension to applied force for small extensions receives its mathematical expression in Hooke's law.

EXERCISES ON 4.2

1. Write down the equation of the tangent to the curve $y = x^2$ at the point $x = 1$. Using the tangent as a linear approximation to the function $f(x) = x^2$ near $x = 1$, obtain approximate values for $f(1.1)$, $f(1.01)$ and $f(1.001)$. How small must δ be in order that the approximation to $f(1 + \delta)$ obtained in this way (with $\delta > 0$) is accurate to 0.1%?

2. Examine the behaviour under scaling about the origin of the functions $f_1(x) = x|x|$, and $f_2(x) = |x(x + 1)|$, and explain how this relates to the differentiability or non-differentiability of the functions. If you have access to software which will plot graphs and which has a 'zoom' capability, use this to investigate scaling.

3. Provide further examples of physical laws which may be expressed in terms of **linear** relationships between measurable quantities. In each case, identify any limitations in the range of applicability of linearity — is linearity valid over the whole range, or just a local approximation?

4.3 Speed and velocity

Perhaps our most direct experience of the idea of derivative as rate of change may be while travelling along a motorway at 70 miles per hour, or faster still on a roller-coaster at a fair. We take it for granted that at any instant of time t we are travelling at some definite speed, and that our speed can change as the value of t changes.

Let us see how to describe mathematically a moving object such as a car or a roller-coaster. Let s be the distance travelled by the object in time t. Taking $t = 0$ to be zero time at which the object begins its journey, s will be defined for $t \geq 0$, and we can regard s as a function of t. The object need not necessarily move along a straight line, but in any case $s(t)$ will measure the total distance travelled by the object between time 0 and time t. Certainly $s(t)$ can only increase or remain constant as t increases, and if $s(t)$ remains constant then the object must be at rest.

The simplest case, apart from that of a body at rest (in which case s is a constant function), is that in which s depends **linearly** on t, say $s = a + ut$, where a and u are constants. In this case, between times t_1 and t_2 the object moves a distance $(a + ut_2) - (a + ut_1) = u(t_2 - t_1)$, and the value of u, the distance travelled per unit time, is the **speed** of the object, and remains constant.

More generally, if s is not necessarily linear in t, we can define the instantaneous speed to be the rate of change of distance with respect to time. This (instantaneous) speed is just $s'(t)$, the derivative with respect to time of the distance travelled. A more common notation is to write $\dot{s}(t)$ rather than $s'(t)$. Thus $\dot{s}$ is the result of a limit in which we consider the ratio of distance travelled to time taken, over smaller and smaller time intervals. We have already introduced this kind of derivative earlier in the chapter. Graphically, if we plot distance travelled against time t, then the speed is the slope of the graph at a given value of t.

It is often helpful to use coordinates in describing the motion of objects. Consider, for example, a point P moving along the x-axis, and let $x = x(t)$ be the coordinate of the point as a function of time t. Here $x(t)$ is the **displacement** of the point P from the origin O as a function of time. Note that, in contrast to the situation with respect to distance travelled, the displacement $x(t)$ may be positive or negative, and may increase, decrease, or remain constant as t increases.

The rate of change $\dot{x}(t)$ of displacement with respect to time is called the (instantaneous) **velocity** of the point P. If $\dot{x}$ is positive, then P is moving (instantaneously) to the right, and if $\dot{x}$ is negative then P is moving to the left. If $\dot{x}$ has the value zero for some value of t, then we say that the point P is **instantaneously** at rest. In Section 6.2 I shall discuss further the role of the sign of the derivative in determining the increase and decrease of functions. Of course 'instantaneous' here is the operative word. If your roller-coaster has slowed down from 90 miles per hour to arrive at rest at time t, but is about to speed up again to the same speed, but in the opposite direction, then you probably do not feel very much at rest; and if you run your bicycle into a brick wall or other obstacle, then at the moment of impact you may not regard yourself as being at rest either.

Notice once more that velocity has the graphical interpretation of a gradient if we plot displacement x against time t. Velocity and speed are, in fact, quite closely connected, and you will probably be able to convince yourself that speed is just $|\dot{x}|$, the absolute value of velocity.

For a point moving in more than one dimension, more than one coordinate is needed to describe the motion. For example, if a point P moves in three-dimensional space, the position of P may be described at any instant of time by the three coordinates x, y, z with respect to some set of Cartesian axes Ox, Oy, Oz. Each of these coordinates may depend on time t, and we then have **three** functions $x(t), y(t), z(t)$ which together describe the motion. For each value of t, these three coordinates may be regarded as components of a vector $(x(t), y(t), z(t))$, called the **position vector** of the point P, at time t. Usually the position vector is denoted by $\mathbf{r}(t)$; so we have $\mathbf{r}(t) = (x(t), y(t), z(t))$. We can now take the derivative of each of the three coordinate functions. We may then interpret $\dot{x}$, $\dot{y}$ and $\dot{z}$ as the respective components of the velocity of the point P in the x, y and z directions. The vector $\dot{\mathbf{r}}(t)$ having components $(\dot{x}, \dot{y}, \dot{z})$ is called the **velocity vector** of P at time t. In this case, for the point P to be instantaneously at rest requires **all three** components $\dot{x}$, $\dot{y}$ and $\dot{z}$ of the velocity vector to be zero, at the same value of t.

The relationship between speed and velocity has its generalization to motion in three dimensions, though a full treatment requires some familiarity with basic ideas in vector algebra. Just as in one dimension speed is the absolute value of velocity, so in three dimensions speed is the **magnitude**, or **length** of the velocity vector. Since a three-dimensional vector having components (X, Y, Z) has magnitude (by Pythagoras's Theorem) $\sqrt{X^2 + Y^2 + Z^2}$, we have speed $= |\dot{\mathbf{r}}(t)| = \sqrt{(\dot{x})^2 + (\dot{y})^2 + (\dot{z})^2}$.

The later development of these ideas belongs more properly to the field of kinematics and dynamics, and I shall not pursue the subject further at this stage. The notion of derivative plays a central role in the very foundations of classical dynamics, and in later chapters we shall see some further applications of basic calculus in this area.

EXERCISES ON 4.3

1. A point P moves along the x-axis during the time interval from $t = 0$ to $t = 2$, the x coordinate of the point being given as a function of t by $x(t) = 2t - t^2$ ($0 \leq t \leq 2$). Describe in words the motion of P from $t = 0$ to $t = 2$. For $0 \leq t \leq 2$, sketch graphs showing
   (a) the displacement (i.e. x coordinate) of P, as a function of t;
   (b) the distance travelled by P, up to time t;
   (c) the velocity of P, as a function of t; and
   (d) the speed of P, as a function of t.
   Express the speed as a function of distance travelled, for values of t in the interval $0 \leq t \leq 1$, and hence or otherwise sketch a graph of speed against distance travelled.
2. A particle moves in two dimensions, with Cartesian coordinates given as functions of time by $x = \cos at$, $y = \sin at$, where a is a constant. Describe the path of the particle, and show that the particle moves along this path with constant speed.

Summary

The basic idea of derivative which I have tried to emphasize in this chapter is that of a rate of change of a function. This rate of change is obtained by dividing the change in output by change in input, in the limit where the change in input approaches zero. Formally, the derivative is given by the limit

$$f'(x) = \lim_{h \to 0} \frac{f(x + h) - f(x)}{h}$$

and if f is any differentiable function then f' is a corresponding function having the same domain as f. If the limit exists as h tends to zero through positive values, we say that f is right differentiable, and the limit is called the right derivative of f, at x. If the limit exists as h tends to zero through negative values, we refer to the left derivative. For f to be differentiable at x, **both** right and left derivative must exist and be equal. If this is the case at all x in the domain of f then f is indeed a differentiable function.

Graphically, the derivative of a function f at $x = x_0$ is the gradient of the graph $y = f(x)$ at the point (x_0, y_0), where $y_0 = f(x_0)$, and the equation of the tangent to the graph at (x_0, y_0) is then

$$y - y_0 = f'(x_0)(x - x_0)$$

The tangent may be regarded as the best linear approximation to the graph near the point (x_0, y_0), and we can say that the graph of any differentiable function looks locally like a straight line.

In dynamics, both the speed and velocity of a particle are examples of derivatives; the speed is the rate of change of distance travelled, with respect to time, and the velocity is the rate of change of displacement (or of position vector, in the case of a particle moving in two or three dimensions) with respect to time.

FURTHER EXERCISES

1. Explain why a function which is **differentiable** at $x = x_0$ must also be continuous at $x = x_0$. Give an example of a function which is continuous at a point, but not differentiable there.
2. The function $f(x) = \cos x$ is known to satisfy the trigonometric identity $\cos A - \cos B = 2\sin((B - A)/2)\sin((B + A)/2)$. Use this fact, together with the result $\lim_{\theta \to 0} \sin \theta/\theta = 1$, to show that $f'(x) = -\sin x$.
3. Functions f and g are related by $g(x) = xf(x)$. Show that

$$\frac{g(x + h) - g(x)}{h} = x\left\{\frac{f(x + h) - f(x)}{h}\right\} + f(x + h)$$

   Hence show that $g'(x) = xf'(x) + f(x)$. Use this result to show that the derivative of x^n is nx^{n-1}, for any positive integer n.
4. Use the formula $u'(t) = \lim_{h \to 0}[u(t + h) - u(t)]/h$ to evaluate the derivative of the functions $t^2 + 2t$, $1/t^2$.
5. Obtain the equation of the tangent to the curve $y = x^2 + 2x$, at the point $(1,3)$, and the tangent to the curve $y = 1/x^2$ at the point $(2, \frac{1}{4})$.
6. Differentiation of $\sqrt{x}$: for any $x > 0$, let h satisfy $|h| < x$. Verify that $1 + (h/2x) - (h^2/2x^2)$, $1 + (h/x)$, and $1 + (h/2x)$ are all positive. Verify also the inequalities $[1 + (h/2x) - (h^2/2x^2)]^2 < [1 + (h/x)]$ and $[1 + (h/x)] < [1 + (h/2x)]^2$, and hence show that

$$1 + \frac{h}{2x} - \frac{h^2}{2x^2} < \sqrt{1 + \frac{h}{x}} < 1 + \frac{h}{2x}$$

   Use this result, together with the identity $f'(x) = \lim_{h \to 0}[f(x + h) - f(x)]/h$ with $f(x) = \sqrt{x}$, to show that $f'(x) = 1/2\sqrt{x}$ in this case.

5 • Differentiation

The derivative of a function is a function. Given any differentiable function f, there will be a corresponding function f', the derivative of f.

A function is a rule for sending real numbers to real numbers. This rule may be interpreted in different ways but in terms of the graph, a function f is defined by the rule which sends the x coordinate of any point of the graph into the y coordinate

— that is why we can write $y = f(x)$.

The function f' is then defined by the rule which sends the x coordinate of any point of the graph into the **gradient** at that point of the graph

— that is why we can write $y' = f'(x)$, where y' now stands for the gradient or slope at the point (x, y).

This correspondence between a function and its derivative allows us to think in terms of an operator, or mapping, or transformation, which sends functions into their derivatives.

The operation of sending functions to their derivatives is called differentiation. It is so important that there is a special symbol for it. Or rather a variety of special symbols, since there is no commonly agreed notation which is appropriate to all situations. Perhaps the simplest notation for differentiation is to use the symbol D. Then, if f is a (differentiable) function and f' the derivative of f, we write $f' = Df$. Here D stands for 'differentiate', and f' is derived from f by the operator 'differentiate'. If we think of f as defined by the rule $x \to f(x)$, then we can write $f'(x) = (Df)(x)$. An alternative and more common notation, and one which reflects more closely the interpretation of derivative as rate of change, is to write d/dx instead of D. The operator of differentiation is then written as d/dx (and pronounced 'dee by dee x'), and the derivative is written as either $f' = df/dx$, or $f'(x) = df(x)/dx$ if we wish to emphasize the functional dependence of f' on a 'variable' x. Of course with $t \to f(t)$ we should write $f' = df/dt$ or $f'(t) = df(t)/dt$. [This commonly accepted notation, however, carries with it a risk — it is **not** advisable to write, for example, $df(1)/dx$ or $df(1)/dt$ for $f'(1)$, the derivative at 1. This is far too easily confused with the derivative of a constant function and it is better, with d/dx notation, to write something along the lines of $df/dx\big|_{x=1}$].

In this chapter we shall look at the properties of differentiation as an operator. In particular, we shall see how the operation of differentiation combines with other operations on functions, such as addition, multiplication and composition. We start with **linearity**. The fact that differentiation is a **linear operator** is fundamental to a number of branches of calculus including, as we shall see later, the theory of integration and the solution of differential equations. In Section 5.1 we shall see what the linearity of differentiation means. After that, we shall look at functions in combination, and see how to differentiate products, powers, quotients and compositions, as well as infinite series. As a result of this work, you should know far more about the process of differentiation at the end of the chapter than you did at the beginning.

5.1 Differentiation as a linear operator

In Chapter 4, we saw how to write the derivative as a limit. For a function $f(x)$, writing the operator of differentiation as d/dx, the formal expression of derivative as a limit becomes

$$\frac{d}{dx}f(x) = \lim_{h \to 0} \frac{f(x+h) - f(x)}{h} \tag{5.1}$$

We have already used equation (5.1), or its equivalent, to show that

$$\frac{d}{dx}x^2 = 2x, \quad \frac{d}{dx}\sin x = \cos x, \quad \frac{d}{dx}\cos x = -\sin x$$

and most elementary examples of differentiation can be carried out using equation (5.1).

One of the most fundamental principles of the theory of limits is that the limit of the sum of two functions is the same as the sum of the limits. Put another way, we can say that the respective actions of adding and of taking limits commute. This means that if you add two functions and then take a limit, you arrive at the same result as would be obtained by **first** taking the limit of each of the two functions, and **then** adding these two limits. We shall not attempt here to justify this property of the limiting process, for which the interested reader may consult any elementary text in analysis. However, it is not difficult to deduce an important **consequence** of the principle that the addition of functions and the taking of limits commute. Let $S(x)$ be the sum of two functions $f(x)$ and $g(x)$. Then we have $S(x) = f(x) + g(x)$, and for the derivative of the function S we have

$$\begin{aligned}
\frac{d}{dx}S(x) &= \lim_{h \to 0} \frac{S(x+h) - S(x)}{h} \\
&= \lim_{h \to 0} \frac{(f(x+h) + g(x+h)) - (f(x) + g(x))}{h} \\
&= \lim_{h \to 0} \left\{ \frac{f(x+h) - f(x)}{h} + \frac{g(x+h) - g(x)}{h} \right\} \\
&= \lim_{h \to 0} \frac{f(x+h) - f(x)}{h} + \lim_{h \to 0} \frac{g(x+h) - g(x)}{h} \\
&= \frac{d}{dx}f(x) + \frac{d}{dx}g(x)
\end{aligned}$$

Thus for the sum of two functions we find

$$\frac{d}{dx}(f(x) + g(x)) = \frac{d}{dx}f(x) + \frac{d}{dx}g(x) \tag{5.2}$$

To differentiate the sum of two functions, just differentiate each function separately, and add the two derivatives.

Although I have expressed the argument in formal mathematical language, I hope the basic idea is clear. Since addition of functions and taking limits commute, and since the derivative is in fact a limit, it follows that the addition of functions and taking derivatives commute. So addition commutes with differentiation — whether you add two functions and then differentiate, or differentiate the functions and then add, you will arrive at the same result. Given the derivatives of the functions x^2, $\sin x$ and $\cos x$, we can use equation (5.2) to deduce, for example, that

$$\frac{\mathrm{d}}{\mathrm{d}x}(x^2 + \sin x) = 2x + \cos x, \quad \frac{\mathrm{d}}{\mathrm{d}x}(x^2 + \cos x) = 2x - \sin x$$

and

$$\frac{\mathrm{d}}{\mathrm{d}x}(\sin x + \cos x) = \cos x - \sin x$$

An implicit assumption underlying equation (5.2) is that both functions f and g must be differentiable at x. If f and g are both differentiable functions, that is functions which can be differentiated at each point of their respective domains, then x has to be in the domains of both f and g.

✸ *Example 1*

Consider the function $x^2 + 1/x$. We have already seen that both functions x^2 and $1/x$ are differentiable functions. The function $1/x$ is not differentiable at $x = 0$, but qualifies as a differentiable function because $x = 0$ is not in the domain of the function $1/x$. For $x \neq 0$ one has

$$\frac{\mathrm{d}}{\mathrm{d}x}\frac{1}{x} = -\frac{1}{x^2}$$

The result

$$\frac{\mathrm{d}}{\mathrm{d}x}\left(x^2 + \frac{1}{x}\right) = 2x - \frac{1}{x^2}$$

holds for all x except at $x = 0$. Since $x = 0$ is not in the domain of the function $x^2 + (1/x)$, this function is again a differentiable function.

✸ *Example 2*

Consider the function $x^2 + |x|$. For $x \neq 0$, we have

$$\frac{\mathrm{d}}{\mathrm{d}x}|x| = 1, \qquad x > 0$$
$$= -1, \qquad x < 0$$

Hence for $x \neq 0$

$$\frac{\mathrm{d}}{\mathrm{d}x}(x^2 + |x|) = 2x + 1, \qquad x > 0$$
$$= 2x - 1, \qquad x < 0$$

However, $x^2 + |x|$ is not a differentiable function. The function is not differentiable at $x = 0$, which **is** in the domain.

◈ *Example 3*

Let $S(x) = \sqrt{x} + \sqrt{-x}$. The respective domains of the functions $\sqrt{x}$ and $\sqrt{-x}$ are the intervals $[0, \infty)$ and $(-\infty, 0]$. Since these two intervals intersect only at the single point $x = 0$, the domain of $S(x)$ contains only one point. To be differentiable at x, the function S would need in its domain at least an open interval containing x. Hence $S(x)$ is not in this case a differentiable function.

A second operation which commutes with taking limits is that of multiplication by a real number c. The limit of c times a function is the same as c times the limit of the function. Coupled with the expression in equation (5.1) for derivative as a limit, this implies, for any real number c, that

$$\frac{d}{dx}(cf(x)) = c\frac{d}{dx}f(x)$$

In other words, d/dx commutes with multiplication by c.

The two rules of differentiation which we have discovered so far may be combined in the single equation

$$\frac{d}{dx}(Af(x) + Bg(x)) = A\frac{d}{dx}f(x) + B\frac{d}{dx}g(x) \tag{5.3}$$

where f and g are two functions, assumed differentiable at x, and A and B are any real numbers. To verify equation (5.3), simply note that

$$\frac{d}{dx}(Af + Bg) = \frac{d}{dx}(Af) + \frac{d}{dx}(Bg) \quad \text{(sum of functions)}$$

$$= A\frac{df}{dx} + B\frac{dg}{dx} \quad \text{(multiplication by real number, applied twice)}$$

Of course the original two rules may be recovered from equation (5.3) by considering in turn the special cases $A = B = 1$ and $A = c$, $B = 0$.

Given two functions f and g, a function such as $Af + Bg$, where A and B are real numbers (we sometimes say A and B are **constants**, to distinguish from similar expressions where A and B are themselves functions of x), is referred to as a **linear combination** of f and g. Equation (5.3) is a rule for differentiation of a linear combination of functions. Of course, by repeated application of this rule we can differentiate a linear combination of any finite number of functions, and deduce for example that

$$\frac{d}{dx}\left(\sum_{k=1}^{n} A_k f_k(x)\right) = \sum_{k=1}^{n} A_k \frac{d}{dx}f_k(x) \tag{5.4}$$

where $f_1, f_2, ..., f_n$ are functions differentiable at x, and $A_1, A_2, ..., A_n$ are constants. In Section 5.3 we shall see how to differentiate a linear combination of infinitely many functions.

Equations (5.3) and (5.4) express the fact that differentiation is a **linear operator**. An operator is said to be linear if you can operate on any linear combination by operating on each function separately, and then take the linear combination. (Differentiation commutes with taking linear combinations.) The linearity of differentiation is a fundamental property not shared by all operations on functions. For example, the operation which says 'square the function' is non-linear, because

$$(Af(x) + Bg(x))^2 \neq A(f(x))^2 + B(g(x))^2$$

On the other hand, the operator 'multiply by x^2' is an example of a linear operator, because

$$x^2(Af(x) + Bg(x)) = A(x^2 f(x)) + B(x^2 g(x))$$

Linear operators such as differentiation and multiplication by functions of x will play an important role in our discussion of differential equations in Chapter 10, where we shall make a crucial distinction between 'linear' and 'non-linear' equations.

● *Example 4*

A function f is defined by $f(t) = 2t^2 + (1/t) + 3\cos t$. In this case, differentiation may be written as d/dt or D, and we have

$$\frac{df(t)}{dt} = 4t - \frac{1}{t^2} - 3\sin t$$

or

$$(Df)(t) = 4t - \frac{1}{t^2} - 3\sin t$$

Note f is a differentiable function, with domain $\mathbb{R}\backslash\{0\}$.

EXERCISES ON 5.1

1. An operation L sending functions to functions is defined by the rule $(Lf)(x) = xf(2x)$. (For example, L sends the function $f(x) = \sin x$ into the function $f(x) = x \sin(2x)$.) Verify that L defines a linear operator. Provide other examples of linear operators. Why do we use the word 'linear' in describing linear operators?

2. A function f, not the zero function $f(x) = 0$ for all x, is said to be an **eigenfunction** of a linear operator L, with eigenvalue λ, if $(Lf)(x) = \lambda f(x)$ for all x in the domain of f, where λ is a given real number. If L is defined by $(Lf)(x) = f(-x)$, show that $\lambda = +1$ and $\lambda = -1$ are both eigenvalues of L, and give examples of functions which are eigenfunctions of L, corresponding to these eigenvalues.

3. Differentiate each of the functions $\cos x - 3\sin x$, $1 + 2t + 3t^2$ and $x|x| + 1/x$, using the notations d/dx, d/dt, D, or f' as appropriate.

5.2 Products, quotients, powers, ...

Having seen the part played by linearity in the differentiation of sums, and more generally linear combinations, of functions, the natural next step is to look at the way differentiation behaves with respect to **multiplication** of functions. We can again start with equation (5.1) which expresses the derivative as a limit, observing on this occasion the principle, which we shall state rather than prove, that the limit of a product of two functions is the same as the product of the limits.

Given two functions f and g, assumed differentiable at x, we can apply equation (5.1) to evaluate the derivative of the product fg, giving

$$\frac{d}{dx}(f(x)g(x)) = \lim_{h \to 0} \frac{f(x+h)g(x+h) - f(x)g(x)}{h} \qquad (5.5)$$

Really, we would very much like to express the numerator on the right-hand side, which measures the change in the product fg as x is changed to $x+h$, in terms of the respective changes of the function f and of the function g. A convenient way to do this is to make use of the identity

$$\begin{aligned} f(x+h)g(x+h) &- f(x)g(x) \\ &= \{f(x+h) - f(x)\}\{g(x+h) - g(x)\} \\ &\quad + f(x)(g(x+h) - g(x)) + g(x)(f(x+h) - f(x)) \end{aligned}$$

This identity, which is less formidable than it looks, follows from the equation

$$FG - fg = (F-f)(G-g) + f(G-g) + g(F-f).$$

which may readily be verified using a little algebra.

Now substituting for the numerator on the right-hand side of equation (5.5) and rearranging the terms a little, we have

$$\begin{aligned} \frac{d}{dx}(f(x)g(x)) &= \lim_{h \to 0} h\left\{\frac{f(x+h) - f(x)}{h}\right\}\left\{\frac{g(x+h) - g(x)}{h}\right\} \\ &\quad + \lim_{h \to 0} f(x)\left(\frac{g(x+h) - g(x)}{h}\right) \\ &\quad + \lim_{h \to 0} g(x)\left(\frac{f(x+h) - f(x)}{h}\right) \end{aligned} \qquad (5.6)$$

We know that $[f(x+h) - f(x)]/h$ and $[g(x+h) - g(x)]/h$ converge respectively to the limits $d(f(x))/dx$ and $d(g(x))/dx$. Hence the first limit on the right-hand side of equation (5.6), with the factor h multiplying the product, converges to zero. Taking the limit $h \to 0$ in the remaining two limits in equation (5.6) we have

$$\frac{d}{dx}(f(x)g(x)) = f(x)\frac{dg(x)}{dx} + g(x)\frac{df(x)}{dx} \qquad (5.7)$$

This is the rule for differentiating the product of two functions. To differentiate a product, there will be two contributions to add together. In one contribution, only the function g of the product is differentiated, the other factor f being unaltered; whereas in the other contribution it is f which is differentiated while g is unaffected by the differentiation.

The product rule for differentiation very considerably enlarges the class of functions which we are able to differentiate without having constant recourse to differentiation from first principles. The differentiation of powers may be regarded as a simple consequence of the product rule. Starting only from the knowledge that $d(x)/dx = 1$, we have

$$\frac{d}{dx}x^2 = \frac{d}{dx}(x.x) = x\frac{d}{dx}x + x\frac{d}{dx}x = 2x$$

$$\frac{d}{dx}x^3 = \frac{d}{dx}(x^2.x) = x^2\frac{d}{dx}x + x\frac{d}{dx}x^2 = x^2.1 + x.2x = 3x^2$$

$$\frac{d}{dx}x^4 = \frac{d}{dx}(x^3.x) = ...$$

giving in general the result

$$\frac{d}{dx}x^n = nx^{n-1}$$

The corresponding result for the derivative of the nth power of a function

$$\frac{d}{dx}(f^n) = nf^{n-1}\frac{df}{dx}$$

may be obtained in a similar way, and is obtained in question 1 of Exercises on 5.2.

As a further consequence of the product rule for differentiation the rule for differentiating the ratio or quotient of two functions may be deduced. Given two functions f and g, the quotient function q is defined by $q(x) = f(x)/g(x)$. The domain of q consists of all x which belong to both the domain of f and the domain of g, and for which $g(x) \neq 0$. It may be verified from first principles that q is differentiable wherever both f and g are differentiable, provided $g(x) \neq 0$. To evaluate the derivative of q, we may regard f as a product. Thus $f = g.q$, so that by the product rule we have

$$\frac{df}{dx} = g\frac{dq}{dx} + q\frac{dg}{dx}$$

Solving for dq/dx now gives

$$\frac{dq}{dx} = \frac{\frac{df}{dx} - q\frac{dg}{dx}}{g} = \frac{g\frac{df}{dx} - gq\frac{dg}{dx}}{g^2}$$

Substituting $gq = f$ in the numerator now yields

$$\frac{dq}{dx} = \frac{g\frac{df}{dx} - f\frac{dg}{dx}}{g^2}$$

Thus we have the so-called **quotient rule** for differentiation:

$$\frac{d}{dx}\left(\frac{f(x)}{g(x)}\right) = \frac{g(x)\frac{d}{dx}f(x) - f(x)\frac{d}{dx}g(x)}{(g(x))^2} \tag{5.8}$$

As an important special case, with $f(x)$ the constant function $f(x) = 1$, we find

$$\frac{d}{dx}\left(\frac{1}{g(x)}\right) = -\frac{1}{(g(x))^2}\frac{d}{dx}g(x) \tag{5.9}$$

◈ Example 5

$\frac{d}{dx}(x^4 \cos x) = x^4 \frac{d}{dx}\cos x + \cos x \frac{d}{dx}x^4 = -x^4 \sin x + 4x^3 \cos x$

◈ Example 6

The quotient rule enables us to differentiate any **rational** function of x. A rational function of x is a ratio of two polynomials, where a polynomial of degree n is an expression of the form $a_0 + a_1 x + a_2 x^2 + ... + a_n x^n$, with $a_0, a_1, ..., a_n$ all real numbers and $a_n \neq 0$. As an example of the differentiation of a rational function, let $f(x) = (1 + x + x^2)/(2 + x^2)$. Then

$$\frac{df}{dx} = \frac{(2+x^2)\frac{d}{dx}(1+x+x^2) - (1+x+x^2)\frac{d}{dx}(2+x^2)}{(2+x^2)^2}$$

$$= \frac{(2+x^2)(1+2x) - (1+x+x^2).2x}{(2+x^2)^2}$$

which simplifies to

$$\frac{d}{dx}\left(\frac{1+x+x^2}{2+x^2}\right) = \frac{2+2x-x^2}{(2+x^2)^2}$$

EXERCISES ON 5.2

1. Assuming the product rule for differentiation, obtain the derivative of the product of **three** functions, in the form

$$\frac{d}{dx}(fgh) = \frac{df}{dx}.g.h + f.\frac{dg}{dx}.h + f.g.\frac{dh}{dx}$$

How does this result generalize to the derivative of the product of n functions $f_1, f_2, ..., f_n$? Deduce the formula

$$\frac{d}{dx}(f^n) = nf^{n-1}\frac{df}{dx}$$

Assuming the quotient formula for differentiation, show that this formula for the derivative of the nth power of a function applies also if n is a **negative** integer. Write down expressions for the derivatives

$$\frac{d}{dx}(1+x^2)^n, \frac{d}{dx}(1+\cos^2 x)^n$$

2. A spherical balloon is inflated, such that the radius of the balloon is given as a function of time t by $R = R(t)$. It is known that, at time $t = t_1$, $R = R_1$ and

$dR/dt = U_1$. Express in terms of R_1 and U_1 the rate of change with respect to time, at $t = t_1$, of (a) the surface area, and (b) the volume of the balloon.

3. Evaluate the derivative

$$\frac{d}{dt}\{t^2(1 + t^2)^2\}$$

by two different methods, and verify that you obtain the same result in each case.

4. Use linearity of differentiation, and the product rule, to evaluate

$$\frac{d}{dx}(a \sin x + bx \cos x + cx^2 \sin x)$$

where a, b and c are constants. Hence find a function $F(x)$ which satisfies

$$\frac{d}{dx}F(x) = x^2 \cos x$$

5.3 ... and power series

Polynomials and rational functions are each examples of types of functions defined in such a way as will permit exact evaluations. Once a value of x is given, the corresponding value of $f(x)$ may be calculated exactly, by means of a finite number of basic operations (multiplication, division, addition, subtraction). Many of the most important functions in calculus, however, cannot be expressed in such terms. These functions, a number of which will be considered in more detail in Chapter 6, include the trigonometric, exponential and hyperbolic functions, logarithmic functions, and numerous more exotic functions such as the Bessel functions. Such functions require an infinite number of basic operations. More precisely, they can all be expressed in terms of power series.

A power series is an infinite series of the form

$$a_0 + a_1 x + a_2 x^2 + \dots$$

where the coefficients $a_0, a_1, a_2, \dots$ are real numbers. The sum of the series up to and including power x^N is then

$$S_N(x) = a_0 + a_1 x + a_2 x^2 + \dots + a_N x^N$$

For given x, if $S_N(x)$ approaches a limit as N tends to infinity, we shall say that the series is convergent (for this value of N). Otherwise, the series is said to be divergent. For a convergent series, this limiting value of $S_N(x)$ is defined to be the sum of the infinite series, and written as $\sum_0^\infty a_n x^n$, or simply $a_0 + a_1 x + a_2 x^2 + \dots$. Denoting by $S(x)$ the sum of a convergent infinite series, we therefore have

$$S(x) = \lim_{N \to \infty} S_N(x) = \sum_0^\infty a_n x^n \tag{5.10}$$

Since for a convergent series we also have $S(x) = \lim_{N \to \infty} S_{N-1}(x)$, so that $\lim_{N \to \infty} a_N x^N = \lim_{N \to \infty}\{S_N(x) - S_{N-1}(x)\} = S(x) - S(x) = 0$, it follows that the Nth term of a convergent power series tends to zero in the limit $N \to \infty$. Although this is so for every convergent series, it is by no means true that a series for which the Nth term converges to zero will necessarily be convergent.

To know whether $S(x)$ makes sense, for given x, we need to know whether the series is convergent or not. There are many tests which may be applied to series, to determine whether they are convergent. The theory of convergence for power series and for other types of infinite series is part of the subject matter of analysis, and will not be treated in any detail here. No practical convergence test can deal successfully with **every** example of power series — there will always be cases for which any given test will be unable to determine whether the series converges or not. One very useful and straightforward test, which depends for its justification on a comparison with the geometric series, is the so-called **ratio test**. To apply the ratio test to an infinite series $c_0 + c_1 + c_2 + ...$, which need not necessarily be a power series, evaluate the ratio c_{N+1}/c_N of consecutive terms of the series, and suppose that the absolute value of this ratio converges to a limit β as N tends to infinity. Thus $\lim_{N \to \infty} |c_{N+1}/c_N| = \beta$.

The ratio test says that the series $\sum_0^\infty c_N$ is convergent if β is smaller than 1, and divergent if β is greater than 1. In the case $\beta = 1$, the ratio test is inconclusive, and has nothing to say about convergence or divergence.

Applied to a power series $\sum a_n x^n$, we must suppose first of all that the ratio $|a_{N+1}/a_N|$ tends to a limit as N tends to infinity. If we call this limit γ, and note that γ is independent of the value of x, the ratio of consecutive terms of the power series will be $a_{N+1}x/a_N$, which tends in absolute value to $\gamma|x|$. Hence, by the ratio test, our power series will converge if $\gamma|x| < 1$ and diverge if $\gamma|x| > 1$. The number $1/\gamma$, which we shall denote by R, is called the **radius of convergence** of the power series. Thus the radius of convergence is the number R such that the power series will converge for $|x| < R$ and diverge for $|x| > R$. It is the critical number which determines convergence or divergence of the series, dependent on the magnitude of $|x|$. Despite the fact that the ratio test does not apply in general (because $|a_{N+1}/a_N|$ for a general power series need **not** tend to a limit), it can be shown that every power series has a unique radius of convergence R, that is a number such that $|x| < R$ implies convergence and $|x| > R$ implies divergence. Two special cases are $R = 0$ and $R = \infty$. If $R = 0$ the series fails to converge for **any** x, except in the trivial case $x = 0$ for which the series has effectively only a single term. One adopts the convention of radius of convergence $R = \infty$ if the series converges for **all** real values of x — with the ratio test this corresponds to the case $\gamma = 0$ in which the ratio a_{N+1}/a_N converges to zero.

Note that for $|x| = R$, a more detailed analysis of individual cases is needed to determine whether or not the series converges, and indeed one may have, for example, convergence for $x = +R$ and divergence for $x = -R$. Once the radius of convergence R has been established for a given power series $\sum a_n x^n$, either through the use of the ratio test or by other means, a function f can be defined, with domain $(-R, R)$, by

$$f(x) = \sum_{n=0}^{\infty} a_n x^n \qquad (-R < x < R) \tag{5.11}$$

(In the case of infinite radius of convergence, $f(x)$ is defined for all real values of x.)

What kind of function f is defined in this way? In particular, can we differentiate equation (5.11) and thereby obtain an infinite series for $d(f(x))/dx$? Such differentiation might not appear at first sight to be legitimate. The problem is that two separate limiting operations are involved. The first is the limit that has to be taken, according to the definition of the derivative, in differentiating each term $a_n x^n$ of the power series. The second limiting operation is the limit $N \to \infty$ that defines the sum of an infinite power series. If these two limiting operations commute, then we can evaluate $df(x)/dx$ by **first** differentiating each term of the power series and then summing to infinity the resulting differentiated series. In fact, the two limits **do** commute, though to prove that this is so would involve detailed arguments of analysis that would take us far beyond the limits of this book. A power series, with radius of convergence R, can **always** be differentiated for $|x| < R$, and the result of this differentiation is just what we would expect:

$$\frac{df(x)}{dx} = \sum_{n=1}^{\infty} n a_n x^{n-1} \qquad (-R < x < R) \tag{5.12}$$

The fact that the right-hand side of equation (5.12) converges implies that the radius of convergence of the differentiated series must be at least as large as that of the original series; otherwise there would be values of x with $|x| < R$ such that the right-hand side diverged. In fact, one may show that the two radii of convergence are the same — a power series and its derivative have the same radius of convergence.

● Example 7

Consider the power series $1 + 2x + 3x^2 + 4x^3 + \dots$. Starting the counting at $n = 0$, the nth term of this series is $(n+1)x^n$. Hence the ratio of the $(n+1)$th to the nth term is $[(n+2)x^{n+1}]/[(n+1)x^n] = [(n+2)/(n+1)]x$. Since $|[(n+2)/(n+1)]x|$ converges to $|x|$ as n tends to infinity, the ratio test gives convergence for $|x| < 1$ and divergence for $|x| > 1$. In fact, the nth term of the series does not converge to zero for $x = \pm 1$, so that the series must diverge for $x = +1$ and $x = -1$. The radius of convergence of the series is $R = 1$. In fact, we have

$$1 + 2x + 3x^2 + 4x^3 + \dots = \frac{d}{dx}(1 + x + x^2 + x^3 + \dots) = \frac{d}{dx}(1-x)^{-1} = (1-x)^{-2}$$

on summing the geometric series.

Hence $1 + 2x + 3x^2 + 4x^3 + \dots = (1-x)^{-2}$; note how summing this series helps to explain the divergence of the series at $x = +1$, though the divergence at $x = -1$ is not quite so obvious.

● Example 8

In Chapter 7, we shall consider the special properties of the exponential function $\exp(x)$, which is defined by the power series

$$\exp(x) = 1 + x + \frac{x^2}{2!} + \frac{x^3}{3!} + \frac{x^4}{4!} + \dots$$

To determine the radius of convergence of this series, note that the ratio of consecutive terms is $[x^{n+1}/(n+1)!]/(x^n/n!)$, which converges to zero whatever the value of x. Since $0 < 1$, the series converges for all values of x, by the ratio test.

1. Use Example 7 to obtain a power series for the function $(1 - x)^{-3}$, and apply the ratio test to verify that this series is convergent for $|x| < 1$.
2. A function f has power series expansion $f(x) = a_0 + a_1x + a_2x^2 + \dots$. Assuming that it is legitimate to multiply power series together and collect terms containing each power of x, show that the function $(f(x))^2$ has a series expansion $a_0^2 + 2a_0a_1x + (a_1^2 + 2a_0a_2)x^2 + \dots$. Hence obtain a power series expansion, up to and including the x^3 term, for the function $\sqrt{1 + x}$.

5.4 The chain rule and differentiation of inverse functions

In Section 3.2, we saw how to take the composition of a pair of functions f, g. Suppose, now, we are given two functions f, g, both of which we know how to differentiate. How do we differentiate their composition $f \circ g$? It is simplest to start by looking at an example, so let us take, say, $f(x) = \cos x$ and $g(x) = x^2$, in which case the composition is given by $(f \circ g)(x) = \cos(x^2)$. Figure 5.1 shows the successive stages in the evaluation of this composite function.

We know that the derivative of the function x^2 is $2x$, and that the derivative of the function $\cos x$ is $-\sin x$. I would like us to think of these derivatives, for the moment, in a slightly different way, as 'amplification factors'. The derivative $2x$, for example, being a limiting ratio, for small variations, of the change in x^2 divided by the change in x, may be thought of as the factor by which (in this limit) we have to multiply any small change in x to obtain the corresponding small change in x^2. So, for small changes, 'change in x^2' $= 2x \times$ 'change in x'. The factor $2x$ amplifies the change in x to give the change in x^2, though of course if $|2x| < 1$ this corresponds to a reduction rather than an expansion, and the word amplification is not so appropriate. In any case, in passing through the first box, the function x^2 in Fig 5.1, any small change in the input value x has to be multiplied by $2x$ to give the resulting small change in output value x^2. In the same way, in passing through the second box, the cosine function, a second amplification factor, which is minus the sine of the input value, must be multiplied into the change in input value to give the change in output value.

Taking the function $f \circ g$ as a whole, this gives, on multiplying by the successive factors corresponding to the two functions, an amplification which is the **product** $2x \times -\sin(x^2) = -2x\sin(x^2)$. Note here that we do not have $-\sin x$ but $-\sin x^2$,

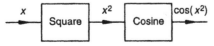

Fig 5.1 Cos (x^2) as a composition.

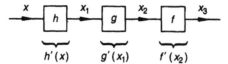

Fig 5.2 Differentiation of a composition.

since not x but x^2 is the input into the cosine box. Since we are interpreting derivatives as amplification factors, this results in the derivative of the composite function being given by

$$\frac{d}{dx}\cos(x^2) = -2x\,\sin(x^2)$$

The general rule for differentiating in this way the composition of two or more functions is as follows. First, draw a flow diagram representing the composition. For each function box, write down an 'amplification factor', which will be the derivative of the function in the box, expressed as a function of the input value to that box. Finally, multiply all amplification factors together. Thus in Fig 5.2, where we differentiate the composition of three functions f, g, h, we have

$$\frac{d}{dx}(f \circ g \circ h) = h'(x)g'(x_1)f'(x_2)$$

where $x_1 = h(x)$, $x_2 = g(x_1) = g(h(x))$, and $x_3 = f(x_2) = f(g(h(x))) = (f \circ g \circ h)(x)$. Hence

$$\frac{d}{dx}(f \circ g \circ h) = h'(x)g'(h(x))f'(g(h(x)))$$

❋ *Example 9*

$d(\cos(x^2))/dx = -2x\sin(x^2)$. With $f(x) = \cos x$ and $g(x) = x^2$, the composition $g \circ f$ is the function $(\cos x)^2$, and

$$\frac{d}{dx}(\cos x)^2 = 2\cos x \times -\sin x = -2\sin x \cos x$$

After a little practice and experience of such examples, you will probably be able to write down the derivatives straight away, without recourse to a flow diagram. You may find it useful on occasions to use the **chain rule**. Suppose you wish to apply the chain rule to evaluate dy/dx, where $y = \cos(x^2)$. Of course you know (or I hope you do!) how to differentiate $\cos x$, but the problem here is that you have x^2 rather than x, which is slightly more difficult. Since x^2 is the problem, get rid of this problem by substituting it away! That is, substitute, say, $t = x^2$. The problem is now broken up into: $y = \cos t$, $t = x^2$, $dy/dx = ?$

Differentiating everything you can now leads to

$$\frac{dy}{dt} = -\sin t, \quad \frac{dt}{dx} = 2x, \quad \frac{dy}{dx} = ?$$

And the **chain rule** says that $dy/dx = (dy/dt).(dt/dx)$. This is easy to remember — almost as though the dts cancelled out, though this is rather a mnemonic than a proof.

Hence $dy/dx = -\sin t \times 2x$ which, on substituting back $t = x^2$ to express everything in terms of x, gives

$$\frac{d}{dx}\cos(x^2) = -2x\sin(x^2)$$

The same result as before!

To differentiate $(\cos x)^2$, just write $y = (\cos x)^2$ and substitute $s = \cos x, y = s^2$. Differentiating everything now gives $ds/dx = -\sin x, dy/ds = 2s$. So, by the chain rule, in this case $dy/dx = (dy/ds).(ds/dx) = -2s\sin x = -2\cos x\sin x$ as before.

✻ Example 10

Let $y = 1/(1 + 3\cos^2 x)$. How do we differentiate this function? The substitution $s = \cos x$ leads to $y = 1/(1 + 3s^2)$. If you are still not sure how to differentiate this function, break the problem up further by substituting $u = 1 + 3s^2, y = 1/u$. Hence the original function $y = 1/(1 + 3\cos^2 x)$ is now described by the **three** equations $y = 1/u$, $u = 1 + 3s^2$ and $s = \cos x$ (or you may prefer to think of y as a composition $f \circ g \circ h$, where $f(x) = 1/x$, $g(x) = 1 + 3x^2$, $h(x) = \cos x$). Differentiating these three equations, we have $dy/du = -1/u^2$, $du/ds = 6s$ and $ds/dx = -\sin x$. Probably you can guess (correctly!) in this case that the chain rule takes the form $dy/dx = (dy/du).(du/ds).(ds/dx)$. Hence the derivative is the product $6s\sin x/u^2$, which on expressing everything as a function of x leads to

$$\frac{d}{dx}\frac{1}{(1 + 3\cos^2 x)} = \frac{6\cos x\sin x}{(1 + 3\cos^2 x)^2}$$

which is the required derivative.

The chain rule can also be used to differentiate inverse functions. As an example of this, we shall evaluate the derivative of the function $\tan^{-1} x$. If $y = \tan^{-1} x$, we have to determine dy/dx; but $y = \tan^{-1} x$ implies $x = \tan y$, which on differentiation gives

$$\frac{dx}{dy} = \frac{d}{dy}\frac{\sin y}{\cos y} = \frac{\cos y\frac{d}{dy}\sin y - \sin y\frac{d}{dy}\cos y}{\cos^2 y} = \frac{\cos^2 y + \sin^2 y}{\cos^2 y}$$

$$= 1 + \frac{\sin^2 y}{\cos^2 y} = 1 + \tan^2 y$$

This is saying the same thing as the identity

$$\frac{d}{dy}\tan y = \sec^2 y = 1 + \tan^2 y$$

Having evaluated dx/dy, it is straightforward to use the chain rule to determine dy/dx. In fact the chain rule gives us $1 = dx/dx = (dx/dy)(dy/dx)$, which leads, as we would expect, to $dy/dx = 1/(dx/dy)$. Hence $dy/dx = 1/(1 + \tan^2 y)$. Since it is generally more convenient to express dy/dx as a function of x rather than of y, we can now substitute back $x = \tan y$ to obtain the derivative

$$\frac{d}{dx} \tan^{-1} x = \frac{1}{1 + x^2}$$

This, then, is the general procedure for differentiating inverse functions. To differentiate the function $f^{-1}(x)$, first write $y = f^{-1}(x)$ together with the equation $x = f(y)$. Next evaluate $dx/dy = f'(y)$, assuming that you know how to differentiate the function f. Then $dy/dx = 1/f'(y)$, and finally you can use either of the equations $y = f^{-1}(x)$ or $x = f(y)$ to substitute back in terms of x.

If you want a (not very memorable) formula to express all this, it is

$$\frac{d}{dx} f^{-1}(x) = 1/f'(f^{-1}(x))$$

but it is usually best to work out each case separately, rather than rely on standard formulae.

● *Example 11*

Let us use these ideas to differentiate the inverse sine function. For the definitions of this and other inverse trigonometric functions, see the examples in Section 3.2. The domain of $\sin^{-1}$ is the interval $-1 \le x \le 1$, and on this interval $y = \sin^{-1} x$ means the same as $x = \sin y$ $(-\pi/2 \le y \le \pi/2)$. For values of y in the open interval $-\pi/2 < y < \pi/2$, that is assuming y is not an endpoint, we have $dx/dy = \cos y$. Hence $dy/dx = 1/\cos y$, which we have to express as a function of x. We know that $x = \sin y$, so that the trigonometric identity $\sin^2 y + \cos^2 y = 1$ gives $\cos^2 y = 1 - \sin^2 y = 1 - x^2$.

Evidently, then, $\cos y = \pm\sqrt{1 - x^2}$. We cannot, however, leave the problem in this indeterminate state. We know that $\sin^{-1} x$ is a well-defined function which has, therefore, a unique derivative rather than two choices from which you take your pick according to the day of the week, or the weather. In fact, the question is easily resolved, since for $-\pi/2 < y < \pi/2$ we always have $\cos y > 0$. Hence $\cos y = +\sqrt{1 - x^2}$ is the only possibility, leading to the result $dy/dx = 1/\cos y = 1/\sqrt{1 - x^2}$. We have, then

$$\frac{d}{dx} \sin^{-1} x = \frac{1}{\sqrt{1 - x^2}}$$

which holds for all x in the interval $-1 < x < 1$.

● *Example 12*

It is required to evaluate the derivative $d[\sin^{-1}(2x - 1)]/dx$. There are, in fact, a number of ways in which this can be done. First of all, we can assume what we already know, namely the derivative of the inverse sine function, and use the chain rule. Then $y = \sin^{-1}(2x - 1)$, with $t = 2x - 1$, may be written in terms of two equations, $y = \sin^{-1} t$ and $t = 2x - 1$, which on differentiation give $dy/dt = 1/\sqrt{1 - t^2}$ and $dt/dx = 2$.

Hence, by the chain rule, $dy/dx = (dy/dt)(dt/dx) = 2/\sqrt{1-t^2}$, which on substituting $t = 2x - 1$ yields $dy/dx = 2/\sqrt{1-(2x-1)^2}$, and on simplification:

$$\frac{d}{dx}\sin^{-1}(2x-1) = \frac{2}{\sqrt{4x-4x^2}} = \frac{1}{\sqrt{x(1-x)}}$$

The differentiation which we have carried out requires $-1 < t < 1$, or $0 < x < 1$, and hence is valid for this interval of values of x.

An alternative approach, and one which does not require prior knowledge of the derivative of the inverse sine function, is to move slightly closer to first principles. We then write

$$y = \sin^{-1}(2x-1) \text{ as } 2x - 1 = \sin y, \text{ or } x = \frac{1}{2}(1 + \sin y)$$

Then $dx/dy = \frac{1}{2}\cos y$, which again must be expressed as a function of x. Since $\sin y = 2x - 1$, we can only have $\cos y = \pm\sqrt{1-\sin^2 y} = \pm\sqrt{1-(2x-1)^2}$. However, as before our definition of the inverse sine function allows only $\cos y > 0$, and we are led once more to $dx/dy = \frac{1}{2}\sqrt{1-(2x-1)^2}$, giving again $dy/dx = 1/\sqrt{x(1-x)}$.

EXERCISES ON 5.4

1. Draw a flow diagram exhibiting the function $\sqrt{1+x^2}$ as a composition of the square root function and the function $1 + x^2$. Use your diagram to explain the result

$$\frac{d}{dx}\sqrt{1+x^2} = \frac{x}{\sqrt{1+x^2}}$$

in terms of amplification factors. Show that the same result can be derived
   (a) by use of the chain rule; or
   (b) by differentiation with respect to x of the equation $y^2 = 1 + x^2$.
2. Use a variety of methods to evaluate the derivative of the function $(\cos(x^2))^2$.
3. Considering carefully the definition in Section 3.2 of the inverse cosine function, show that

$$\frac{d}{dx}\cos^{-1}x = -\frac{1}{\sqrt{1-x^2}}$$

for all x in the interval $-1 < x < 1$. Given also that

$$\frac{d}{dx}\sin^{-1}x = \frac{1}{\sqrt{1-x^2}}$$

what does this result tell you about the derivative of the sum $\sin^{-1}x + \cos^{-1}x$?

Summary

Chapter 5 is mostly about rules for differentiation, though I hope that besides telling you how to apply these rules I have gone some way to explaining some of the fundamental ideas which underlie them. I hope, too, that you have worked through some of the exercises in this chapter, which will encourage you to think about and discuss these ideas. The exercises at the end of the chapter will give you a chance to put what you have learnt into practice. By now you will have methods at your disposal which should enable you to differentiate just about any (differentiable!) function which you are likely to meet. The following summary of the rules of differentiation should help you to do this. Rules should not be applied blindly, and are not a substitute for rational thought! Apply them with care, and think about what you are doing.

- $\frac{d}{dx}(Af + Bg) = A\frac{df}{dx} + B\frac{dg}{dx}$; A, B constants (LINEARITY)

- $\frac{d}{dx}(fg) = f\frac{dg}{dx} + g\frac{df}{dx}$ (PRODUCT RULE)

- $\frac{d}{dx}\left(\frac{f}{g}\right) = \frac{g\frac{df}{dx} - f\frac{dg}{dx}}{g^2}$ (QUOTIENT RULE)

- $\frac{d}{dx}x^n = nx^{n-1}$, n an integer

- $\frac{d}{dx}\left(\sum_0^\infty a_n x^n\right) = \sum_1^\infty na_n x^{n-1}$; $|x| < R =$ radius of convergence (POWER SERIES RULE)

- $\frac{dy}{dx} = \frac{dy}{dt} \cdot \frac{dt}{dx} = \frac{dy}{ds} \cdot \frac{ds}{dt} \cdot \frac{dt}{dx} \ldots$ (CHAIN RULE)

- $y = f^{-1}(x)$, $x = f(y)$, $\frac{dy}{dx} = \frac{1}{\frac{dx}{dy}} = \frac{1}{f'(y)}$ (INVERSE FUNCTION RULE)

- $y = (f(x))^n$, $\frac{dy}{dx} = n(f(x))^{n-1}\frac{df(x)}{dx}$ (POWER OF A FUNCTION RULE)

To supplement these rules for differentiation, it may also be helpful to list at this point a number of derivatives of specific functions. In the case of the derivative of powers of x, so far we have verified the result only for integral powers and one or two special cases such as $x^{1/2}$. The derivative of a general fractional or irrational power of x is best obtained by use of the exponential and logarithmic functions, and I shall defer the general proof until Chapter 7 (see in particular exercise 10 at the end of Chapter 7).

- $\frac{d}{dx}x^c = cx^{c-1}$, for $x > 0$

- $\left.\begin{array}{l}\frac{d}{dx}|x| = 1 \text{ for } x > 0 \\ \quad\quad = -1 \text{ for } x < 0\end{array}\right\}$; $|x|$ is not differentiable at $x = 0$

- $\frac{d}{dx}\sin(ax) = a\cos(ax)$ (a constant)

- $\frac{d}{dx}\cos(ax) = -a\sin(ax)$

- $\frac{d}{dx}\tan(ax) = a\sec^2(ax) = \frac{a}{\cos^2(ax)}$

- $\frac{d}{dx}\tan^{-1}x = \frac{1}{1+x^2}$

- $\frac{d}{dx}\sin^{-1}x = \frac{1}{\sqrt{1-x^2}}$ ($-1 < x < 1$)

- $\frac{d}{dx}\cos^{-1}x = -\frac{1}{\sqrt{1-x^2}}$ ($-1 < x < 1$)

FURTHER EXERCISES

1. Write down the derivative of each of the following functions:

$x^2 + x^4$, $2x + \sqrt{x}$, $x^2 \sin x$, $\sin^2 x$, $\sqrt{x} \cos x$, $\frac{x^2}{1+x^4}$, $\frac{1-x^2}{1+x+x^2}$,

$\frac{\sin x}{1+\cos^2 x}$, $\sin(x^2 + 1)$, $x \cos(x^2 + 1)$, $\frac{\sin 2x}{x}$, $(1 + \cos 2x)^3$, $\sum_1^\infty \frac{x^n}{n(1+n^2)}$, $(\sin 2x)^n$,

$\tan^{-1} 2x$, $\tan^{-1} \sqrt{x}$

2. Two functions, denoted by $c(x)$ and $s(x)$, respectively, are known to satisfy the equations $dc/dx = -s$ and $ds/dx = c$, for all $x \in \mathbb{R}$.

   Prove that c and s satisfy the further equations

$$\frac{d}{dx}\left(\frac{c}{s}\right) + \left(\frac{c}{s}\right)^2 + 1 = 0, \quad \frac{d}{dx}(s.c) = c^2 - s^2,$$

$$\frac{d}{dx}(c^2 - s^2) = -4s.c, \quad \frac{d}{dx}(c^2 + s^2) = 0$$

3. A function f is known to satisfy the equation $df/dx = f$, for all $x \in \mathbb{R}$. Show that

$$\frac{d}{dx}f(2x) = 2f(2x) \text{ and } \frac{d}{dx}(f(x))^2 = 2(f(x))^2$$

   and assuming $f(x) \neq 0$ use the quotient rule to show that

$$\frac{d}{dx}\left\{\frac{f(2x)}{(f(x))^2}\right\} = 0$$

4. Use some of the rules for differentiation to verify the identities

$$\frac{d}{dx}\left(\frac{1}{f^2}\right) = -\frac{2}{f^3}\frac{df}{dx}, \quad \frac{d}{dx}\sin(f) = \frac{df}{dx}\cos(f),$$

$$\frac{d}{dx}\left\{x^2 f\left(\frac{1}{x}\right)\right\} = 2xf\left(\frac{1}{x}\right) - f'\left(\frac{1}{x}\right)$$

5. Determine the radius of convergence R of the power series $x + (x^2/2) + (x^3/3) + (x^4/4) + \ldots$. Denoting by $g(x)$ the sum of this power series, evaluate the derivative of the function g.

6 • Finding Out About Functions

Given a function f, it is of interest to know whether a particular change of input value x will give rise to an increase or a decrease in output value $f(x)$, or indeed whether the output value remains the same. If f is differentiable, such questions can often most easily be answered by considering the derivative df/dx, which is after all a rate of change. In this chapter, we shall see what derivatives have to tell us about the local variation of functions, and what can be deduced about global properties such as the position and nature of any maxima or minima. We shall also examine the role of higher order derivatives in the analysis of functions. To extend our knowledge from first order to higher order derivatives will not only improve our understanding of the local behaviour of functions, but will also be an essential prerequisite for the further analysis of power series and our later treatment of differential equations.

6.1 Derivatives, ... and more derivatives

Once we have begun to see differentiation as a linear operator, which sends functions to their derivatives, it is natural to consider the effect on a function of letting this operator act a number of times. If we call the derivative f' of f the **first** derivative, then we define the **second** derivative of f to be the derivative of the first derivative, the **third** derivative to be the derivative of the second derivative, and so on. So the nth derivative of f is obtained from f by successive applications of the operator d/dx n times. Having called the first derivative f', we can denote the second derivative by f'', the third derivative by f''', and so on. An alternative notation (and one which is certainly more convenient for large values of n) is to write $f^{(n)}$ for the nth derivative of f. We have then

$$f'(x) = \frac{d}{dx} f(x)$$

$$f''(x) = \frac{d}{dx}\left(\frac{d}{dx} f(x)\right) = \left(\frac{d}{dx}\right)^2 f(x)$$

$$f'''(x) = \frac{d}{dx}\left(\frac{d}{dx}\left(\frac{d}{dx} f(x)\right)\right) = \left(\frac{d}{dx}\right)^3 f(x)$$

$$\vdots$$

$$f^{(n)}(x) = \overbrace{\frac{d}{dx}\left(\frac{d}{dx}\left(\frac{d}{dx}\cdots f(x)\right)\cdots\right)}^{n \text{ times}} = \left(\frac{d}{dx}\right)^n f(x)$$

Thus the *n*th derivative $f^{(n)}$ is the result of applying the operator d/dx *n* times to *f* and is also called the *n*th order derivative of *f*. We shall also adopt the common abbreviation d^n/dx^n (pronounced 'dee to the *n* by dee *x* to the *n*') for $(d/dx)^n$, so, for example, the second derivative $f''(x) = (d/dx)^2 f(x)$ will usually be written $(d^2/dx^2)f(x)$ ('dee 2 by dee *x* squared' or 'dee squared by dee *x* squared'). Either $(d^2/dx^2)f$ or d^2f/dx^2 will do. If the function *f* is thought of as a mapping $t \rightarrow f(t)$, then the notation d^2f/dt^2 rather than d^2f/dx^2 is used. A common notation which does not distinguish between different symbols for the input variable is to use *D* for differentiation, and denote the *n*th derivative of *f* simply $D^n f$. Whatever convention is used, it is vital **not** to confuse, for example, $(d^2/dx^2)f$ or $(d/dx)^2 f$ with $((d/dx)f)^2$ or with $(d/dx)(f^2)$. It is the **operator** d/dx which is squared, not the function or its derivative!

● *Example 1*

Differentiation of polynomials: the operator d/dx, applied to a polynomial of degree *n*, results in a polynomial of degree $n - 1$. Since each operation by d/dx reduces by one the degree of the polynomial, it follows that if *f* is a polynomial of degree *n* then for $0 \leq k \leq n$, $(d^k/dx^k)f$ is a polynomial of degree $n - k$. Also $d^k f/dx^k = 0$ for all $k > n$. Later we shall see that polynomials are the **only** functions that satisfy an equation of the form $d^k/dx^k f = 0$ for all $x \in \mathbb{R}$. It is a simple exercise in repeated differentiation to show that $(d^n/dx^n)(x^n) = n!$.

● *Example 2*

Trigonometric functions: it is straightforward to evaluate the higher order derivatives of the functions $\sin x$ and $\cos x$. For example

$$\frac{d}{dx}\sin x = \cos x, \quad \frac{d^2}{dx^2}\sin x = -\sin x, \quad \frac{d^3}{dx^3}\sin x = -\cos x,$$

$$\frac{d^4}{dx^4}\sin x = \sin x, \quad \frac{d^5}{dx^5}\sin x = \cos x$$

and so on. After each four successive derivatives, the pattern repeats itself. This allows us to say, for example, that

$$\frac{d^{4k+1}}{dx^{4k+1}}\sin x = \cos x$$

for any positive integer *k*.

Higher order derivatives of other trigonometric functions may be evaluated using the product and quotient rule. However, it is usually difficult to see any clear pattern emerge, even if the higher order derivative is evaluated at specific points such as $x = \pi$. As an illustration, let $f(x) = \tan x = \sin x / \cos x$. Then

$$f'(x) = \frac{\cos x \frac{d}{dx}\sin x - \sin x \frac{d}{dx}\cos x}{\cos^2 x} = \frac{1}{\cos^2 x}$$

$$f''(x) = -2(\cos x)^{-3}\frac{d}{dx}(\cos x) = \frac{2\sin x}{\cos^3 x}$$

where we have used

$$\frac{d}{dx}(f)^{-2} = -2f^{-3}\frac{df}{dx}$$

The third derivative f''' may be evaluated using the quotient rule, and further derivatives found, but these are increasingly complicated and there is no clear pattern.

◈ Example 3

Let

$$f(t) = (1+t)^{-1}$$

Then

$$\frac{d}{dt}f(t) = -(1+t)^{-2}, \quad \frac{d^2}{dt^2}f(t) = 2(1+t)^{-3}, \quad \frac{d^3}{dt^3}f(t) = -3.2\,(1+t)^{-4}$$

and it is not difficult to see that the derivative of order n is given by

$$f^{(n)}(t) = -n!(1+t)^{-(n+1)}$$

◈ Example 4

Let $f(x) = x|x|$. As a result of problem 4 of Exercises on 4.1, we know that f is a differentiable function and that

$$\frac{d}{dx}f(x) = 2|x|$$

However, $2|x|$ is **not** a differentiable function. In particular, the second derivative $d^2(f)/dx^2$ does not exist at $x = 0$, since this would correspond to $d(2|x|)/dx$, which is undefined at this point, so $f(x) = x|x|$ is an example of a function which is differentiable but not **twice** differentiable. There are other examples of functions for which, at various points, derivatives up to some order k exist but the $(k+1)$th derivative does not. These cases show that, even for differentiable functions, there is no guarantee that higher order derivatives can be defined at all points in their domain.

One of the important applications of higher order derivatives is to power series. Let f be a function which has a power series expansion, convergent within its radius of convergence. Then

$$f(x) = a_0 + a_1 x + a_2 x^2 + a_3 x^3 + \dots \qquad (-R < x < R)$$

We know that such a power series can be differentiated term by term, provided we keep within the prescribed interval $|x| < R$. Moreover, the differentiated series which is convergent with the same radius of convergence R, can again be differentiated to obtain a series for the second derivative of f. This process can be continued to give power series expansions for all higher order derivatives of f, again convergent within the same interval. We can list these series, together with the original series for f, as follows:

$$
\left.
\begin{aligned}
f(x) &= a_0 + a_1 x + a_2 x^2 + a_3 x^3 + \dots \\
f'(x) &= a_1 + 2a_2 x + 3a_3 x^2 + \dots \\
f''(x) &= 2a_2 + 3.2a_3 x + \dots \\
f'''(x) &= 3.2a_3 + \dots \\
\dots, \quad &\text{and so on}
\end{aligned}
\right\}
$$

The situation is particularly simple at $x = 0$, where we have $f(0) = a_0, f'(0) = a_1$, $f''(0) = 2a_2$, $f'''(0) = 3.2a_3$, ..., and in general $f^{(n)}(0) = n!a_n$. Hence, $a_n = f^{(n)}(0)/n!$, and this result enables us to determine all the coefficients in the power series for $f(x)$, given the values of f, f', and all higher order derivatives, at the single point $x = 0$. Summarizing this information, we see that the power series for $f(x)$ may be written as

$$
f(x) = f(0) + x f'(0) + \frac{x^2}{2!} f''(0) + \frac{x^3}{3!} f'''(0) + \dots + \frac{x^n}{n!} f^{(n)}(0) + \dots \tag{6.1}
$$

and is convergent within the radius of convergence R. Of course, not all functions may be expanded in a power series in this way. In the case of the function $f(x) = |x|$, not even the first derivative exists at $x = 0$, and for $f(x) = x|x|$ differentiation breaks down at $x = 0$ for the second order derivative. Even a function $f(x)$ for which $f^{(n)}(0)$ exists for all n may not necessarily have a power series expansion. What has been said is that **if** a function can be represented as a convergent series in powers of x, then the coefficients of this power series can be expressed in terms of the successive derivatives of the function, and the series must look like eqation (6.1). The series on the right-hand side of equation (6.1) is then called the **Taylor series**, or Taylor expansion, of the function about $x = 0$. The Taylor series of a function about $x = 0$ is a series in increasing powers of x. The terms of such a series decrease as the value of $|x|$ is reduced, particularly for larger values of n if the coefficients a_n are not too big, and convergence of the series may be expected to be rapid for $|x|$ very small. Hence the expansion about $x = 0$ is likely to be of the greatest practical use if x is taken close to zero, in which case a good approximation to the sum of the entire series may often be obtained by summing only a relatively small number of terms. If rapid convergence, or a good approximation to $f(x)$ by taking only a finite number of terms, is sought for values of x which are not particularly close to zero, it is often better to attempt to express the function as a series in powers of $(x - c)$ for some non-zero value of the constant c. Such a series, which may be expected to converge more rapidly for x close to c, will be of the form

$$
f(x) = A_0 + A_1(x - c) + A_2(x - c)^2 + A_3(x - c)^3 + \dots
$$

convergent for $|x - c| < r_c$, where r_c is a radius of convergence, depending in general on which value of c is chosen. By very similar arguments to those which we used for the series about $x = 0$, it follows that the coefficients in this series are given by $A_n = f^{(n)}(c)/n!$, so again, to determine the coefficients in such a series, it is necessary only to evaluate f, f' and higher order derivatives at the single point $x = c$. The resulting series in powers of $(x - c)$, called the Taylor series or Taylor expansion of the function about $x = c$, is given by

$$f(x) = f(c) + (x - c)f'(c) + \frac{(x - c)^2}{2!}f''(c) + \frac{(x - c)^3}{3!}f'''(c) + \dots$$

$$+ \frac{(x - c)^n}{n!}f^{(n)}(c) + \dots \tag{6.2}$$

Again, we must not expect all functions to be expressible as convergent power series about $x = c$, but if a function can be written as such a series, then the series must necessarily take the form of equation (6.2). The original Taylor expansion about $x = 0$, given in equation (6.1), corresponds of course to the special case $c = 0$.

For a function to have a power series expansion about a given point c, it is **necessary** that c lies in the domain of the function f, and of all derivatives $f^{(n)}$ for $n \geq 1$. We shall not list here the various **sufficient** conditions for the existence of power series, which depend in the most part on estimates of the size of $|f^{(n)}|$ for large values of n. It is enough, for our purposes, to note that virtually all of the standard functions which are of use in calculus satisfy one or other of these estimates, and hence may be expressed, within the appropriate radius of convergence, as Taylor series.

⊛ *Example 5*

Polynomials again: we have already seen, for a polynomial of degree n, that all derivatives of order higher than n are zero. Hence the coefficients of the Taylor series for a polynomial, whether about $x = 0$ or about a general point $x = c$, are zero from some term onwards. It is not surprising, in view of the special nature of polynomial functions, that in this case the Taylor expansion consists of a finite number of terms only. It is, nevertheless, instructive to note, from equation (6.2), that, for example, the Taylor series about $x = 1$ for the function $f(x) = x^3$ leads to

$$x^3 = 1 + 3(x - 1) + 3(x - 1)^2 + (x - 1)^3$$

⊛ *Example 6*

Trigonometric functions: earlier in the chapter, in Example 2, we looked at higher order derivatives for some of the trigonometric functions. In the case of the sine function, we have, at $x = 0$, $f(0) = 0$, $f'(0) = \cos 0 = 1$, $f''(0) = -\sin 0 = 0$, $f'''(0) = -\cos 0 = -1$, after which the sequence of derivatives repeats itself. Substituting these values into the coefficients of the Taylor series then gives

$$\sin x = 0 + 1.x + 0.x^2 + \frac{(-1)}{3!}x^3 + \dots$$

leading to the series

$$\sin x = x - \frac{x^3}{3!} + \frac{x^5}{5!} - \frac{x^7}{7!} + \dots + \frac{(-1)^n x^{2n+1}}{(2n + 1)!} + \dots \tag{6.3}$$

A similar calculation for the cosine function, or simply direct differentiation of equation (6.3), gives

$$\cos x = 1 - \frac{x^2}{2!} + \frac{x^4}{4!} - \dots + \frac{(-1)^n x^{2n}}{(2n)!} + \dots \tag{6.4}$$

These two equations represent the Taylor expansions for the sine and cosine functions about $x = 0$. In each case, it may be verified by means of the ratio test that there is infinite radius of convergence, so both series converge for all values of x. It would be quite possible to take these two series as the starting points for the respective **definition** of the sine and cosine functions, and to deduce the principal properties of these functions from the series. I have not adopted this procedure in this case, mainly because for most students the first introduction to the sine and cosine functions is through their applications in trigonometry, so that for most of us equations (6.3) and (6.4) are not exactly what we **mean** by sine and cosine. We shall, however, follow this approach to some extent in Chapter 7, when we shall use differentiation and some of the basic ideas of calculus to derive some of the principal trigonometric identities.

In the case of the function $\tan x$, considered in Example 2, there is no simple expression for $f^{(n)}(0)$. In fact, the coefficients in the Taylor expansion about $x = 0$ for this function can be expressed in terms of the so-called Bernoulli numbers B_n.

EXERCISES ON 6.1

1. Evaluate the higher order derivatives up to and including order 3 for the function $f(x) = x/(1+x^2)$, expressing the third derivative in the form $d^3f/dx^3 = (a + bx^2 + cx^4)/(1+x^2)^4$, where a, b and c are constants which should be determined.
2. Assuming the product rule for differentiation, show that the second derivative of the product of two functions f, g is given by

$$D^2(fg) = f(D^2g) + 2(Df)(Dg) + (D^2f)g$$

Write down a similar formula for $D^3(fg)$, of the form

$$D^3(fg) = f(D^3g) + p(Df)(D^2g) + q(D^2f)(Dg) + (D^3f)g$$

where p and q are constants to be determined.

The corresponding result for $D^n(fg)$ is called Leibniz's formula, and is the generalization of the product rule of calculus to the nth derivative of a product. What do you think Leibniz's formula looks like? Use Leibniz's formula or other means to evaluate $D^{10}(x \sin x)$.
3. Another way to extend the product rule is to consider the derivative of the product of three of more functions. Write down an expression for $D(fgh)$, the derivative of the product of three functions.
4. Evaluate

$$\frac{d^n}{dx^n}\{(1-x)^{-1}\}$$

and hence write down the Taylor series about $x = 0$ for the function $(1-x)^{-1}$.

Is this series what you expected? Use the ratio test to determine its radius of convergence.

Without further calculation, write down the Taylor series about $x = 0$ for the function $(1 - x^2)^{-1}$. What is its radius of convergence? Use the coefficients of this series to evaluate the nth derivative at $x = 0$ of the function $(1 - x^2)^{-1}$, (a) if n is even, and (b) if n is odd. Obtain also the Taylor series for the function $(1 - x)^{-1}$ about $x = 2$.

6.2 The rise and fall of functions

Any discussion of the way in which the values of a function $f(x)$ change with variations in the input variable x must start by being rather precise about what it means for a function to be **increasing**, or **decreasing**. Two different ideas are involved. First of all, we may talk of a function being increasing at a point of its domain. This is a **local** idea of what 'increasing' means. Or we may talk of a function that is increasing over an interval. This is more of a global idea of an increasing function. The two definitions can be made precise as follows.

● *Definition I*

A function $f(x)$ is said to be **increasing** at a point x_0 of its domain if there is an open interval $a < x < b$, with x_0 belonging to this interval, such that $f(x) < f(x_0)$ for $a < x < x_0$ and $f(x) > f(x_0)$ for $x_0 < x < b$.

A function $f(x)$ is said to be **increasing on an interval** $a < x < b$ if $f(x_1) < f(x_2)$ for all x_1, x_2 satisfying $a < x_1 < x_2 < b$.

The first part of the definition makes precise what is meant by a function increasing at $x = x_0$. This requires that, for all x near to x_0 (that is, for all $x \in (a, b)$), $f(x)$ must be smaller than $f(x_0)$ if x is to the left of x_0, and larger than $f(x_0)$ if x is to the right. To determine whether a function $f(x)$ is increasing at a given point x_0, you have to compare the values of $f(x)$ with $f(x_0)$, for x close to x_0. On the other hand, for a function to be increasing on an interval (a, b), a comparison has to be made between $f(x_1)$ and $f(x_2)$ for any two points x_1, x_2 in the interval, with $x_1 < x_2$. Notice that both definitions assume that all x in the interval (a, b) belong to the domain of the function.

It is a straightforward exercise in clear thinking to show that any function that is increasing on an interval (a, b) is increasing at each point of the interval. Conversely, though not quite so straightforward it is nevertheless true that a function which is increasing at each point of an interval (a, b) is necessarily increasing on the interval (a, b). However, it is quite possible, and frequently happens, that a function may be increasing at certain points of an interval, but not at others.

The obvious amendments to the definition can be made, changing the directions of the inequalities for f, to establish the notion of a function **decreasing** at a point, or decreasing on an interval. Note that if $f(x)$ is increasing, either at a point or on an interval, then $-f(x)$ is decreasing.

As simple illustrations of these ideas, observe that the function x^3 is increasing on the entire real line (since $x_1 < x_2 \Rightarrow x_1^3 < x_2^3$), and hence increasing at every $x \in \mathbb{R}$. The function x^4, on the other hand, is increasing at every $x > 0$ and

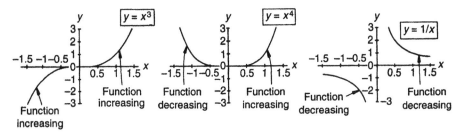

Fig 6.1 Increase and decrease of a function.

decreasing at every $x < 0$; at $x = 0$, x^4 is neither increasing nor decreasing. Hence this function is increasing on the interval $(0, \infty)$ and decreasing on the interval $(-\infty, 0)$. The function $1/x$ is decreasing on $(-\infty, 0)$, and also decreasing on $(0, \infty)$. Notice that despite being decreasing at every point of its domain, this function still has $f(x) < f(x')$ whenever $x < x'$ and $x < 0$, $x' > 0$.

The graphs of these functions are shown in Fig 6.1. Note the graphical interpretation that if $f(x)$ is increasing at x_0 then points of the graph to the right of x_0 are above the level of $f(x_0)$, and points to the left are below the level of $f(x_0)$. If f is increasing on an interval (a, b) then points of the graph are at a higher level as we move from left to right across the interval. One may quite well consider the behaviour of functions with respect to points of increase or decrease without any assumption of differentiability at all. More or less any type of behaviour is possible. A function may have a very wild graph, with no points of increase or decrease at all, or alternatively be increasing at all $x \in \mathbb{R}$, and yet have many points of non-differentiability. (It is a remarkable and quite deep theorem of analysis that any function increasing on an interval must be differentiable at 'almost all' points of the interval, where 'almost all' is to be interpreted in a strict sense using the theory of measures.) However, it is with differentiable functions that the situation becomes clearest, and the following criteria describe what can be said in this case.

Criteria for functions to be increasing

(a) Suppose a function $f(x)$ is differentiable at $x = x_0$. Then $f(x)$ is increasing at x_0 if $f'(x_0) > 0$, and decreasing at x_0 if $f'(x_0) < 0$.

(b) Suppose a function $f(x)$ is differentiable at all points of an interval $a < x < b$, and that $f'(x) > 0$ for all $x \in (a, b)$. Then $f(x)$ is increasing on the interval (a, b).

The proof of (a) and (b) above follows directly from the formal definition of the derivative. If $f'(x_0) > 0$, then we must have $[f(x_0 + h) - f(x_0)]/h > 0$ provided $|h|$ is sufficiently small, so for small enough values of $|h|$, we must have $f(x_0 + h) > f(x_0)$ for $h > 0$ and $f(x_0 + h) < f(x_0)$ for $h < 0$. From Definition 1, it follows that $f(x)$ is then increasing at $x = x_0$. Criterion (b), for a function to be increasing on an interval, is a consequence of (a).

Criterion (b) may actually be weakened in various directions. For example, if the function is known to have positive derivative at all $x \in (a, b)$ except possibly at finitely many points in the interval, where the function is continuous, then again we may conclude that $f(x)$ is increasing on the interval. As a special case, a function differentiable on an interval (a, b), with $f'(x) > 0$ except possibly at finitely many values of x, will be increasing on the interval.

Criteria similar to (a) and (b), with $f' < 0$, hold for functions to be decreasing. If the derivative is **equal to** zero at a point $x = x_0$, then no particular conclusion can be drawn. However, for a function with zero derivative **on an interval**, the situation is dramatically different, as we should expect. In adding the following criterion to our two previous results (a) and (b), its implications for the later development of our subject cannot be too strongly emphasized.

Criterion for functions to be constant

(c) Suppose a function $f(x)$ is differentiable at all points of an interval $a < x < b$, and that $f'(x) = 0$ for all $x \in (a, b)$. Then $f(x)$ is a constant function on the interval (a, b). Conversely, a constant function on the interval (a, b) has zero derivative on the interval.

The formal proof of (c), as well as the slightly improved versions of (b) that I mentioned earlier, depend on a result of analysis called the Mean Value Theorem. The Mean Value Theorem says that if a function $f(x)$ is differentiable at all x in an interval $x_1 < x < x_2$, and continuous at the endpoints x_1, x_2 of the interval, then there exists a point c, with $x_1 < c < x_2$, such that $[f(x_2) - f(x_1)]/(x_2 - x_1) = f'(c)$.

In Fig 6.2, $[f(x_2) - f(x_1)]/(x_2 - x_1)$ is the gradient of the straight line X_1X_2 joining the two points of the graph $y = f(x)$ which correspond to $x = x_1$, $x = x_2$, respectively. The point C, which need not be unique on the interval $x_1 < x < x_2$, is a point $x = c$ of the graph at which the gradient $f'(c)$ of the curve equals the gradient of the chord X_1X_2. The tangent at C is then parallel to X_1X_2. It is not difficult to accept the Mean Value Theorem as a reasonable result, based on the geometrical properties of the graph of a differentiable function. A formal proof may be found in any elementary textbook of analysis.

Given the Mean Value Theorem and applying this result to (c) above, suppose that $f(x)$ is differentiable at all $x \in (a, b)$, with $f'(x) = 0$. For any $x_1, x_2 \in (a, b)$,

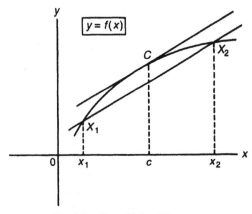

Fig 6.2 Mean Value Theorem.

with $x_1 < x_2$, the theorem allows us to assert the existence of a real number c, such that $x_1 < c < x_2$ and $[f(x_2) - f(x_1)]/(x_2 - x_1) = f'(c)$. However, c lies in the interval (a, b), hence $f'(c) = 0$. It follows that $f(x_1) = f(x_2)$.

We have shown, then, that the function f must assume the same value at **any** two distinct points x_1, x_2 of the interval (a, b). Hence $f(x)$ has the same value at all points of the interval, which is the same as the statement that f is a constant function. The converse result, that any function constant on an interval has zero derivative there, is of course very familiar to us already.

Some who are new to the rigours of mathematical analysis (and some who are not) wonder how it can be necessary to provide formal proofs of results such as (c) at all. Why turn to the Mean Value Theorem when the statement in question is transparently obvious? We should need little convincing that the road from Hull to Cottingham, given that it is of zero gradient along its length, is at a fixed level. Of course every curve with zero slope is the graph of a constant function! This, however, is the penalty for seeking ever more precise definitions of the derivative and related concepts such as continuity. The more successful we are in reaching a higher degree of precision in mathematics, the harder we have to work to be sure that our definitions do indeed guarantee the general conclusions that we would like to draw from them. In the case of the derivative, this often means proceeding from local conditions, which relate to the behaviour of a function in a limit as we approach a given, general, point of an interval, to the global behaviour of the function across the entire interval. To make the argument complete, general results from analysis such as the Mean Value Theorem and the Intermediate Value Theorem play an indispensable role. Though we shall not in this book pursue these matters in detail, the results of analysis underpin much of what we do, and without this underpinning calculus could not be the powerful and universal method which it has developed into.

The following examples apply some of these ideas to specific cases.

Example 7

Let $f(x) = (1 - x)/(1 + x)$. By the quotient rule, we have

$$f'(x) = \frac{(1 + x). - 1 - (1 - x).1}{(1 + x)^2} = -\frac{2}{(1 + x)^2} < 0$$

except at $x = -1$. Hence $f(x)$ is decreasing on each of the two intervals $(-\infty, -1)$ and $(-1, \infty)$. The graph of this function is shown in Fig 6.3. Note that $y \to -1$ as $x \to \pm\infty$. Also $y \to +\infty$ as x approaches -1 from the right, and $y \to -\infty$ as x approaches -1 from the left.

Example 8

Let $f(x) = x/(1 - x^2)$. Here, by the quotient rule, we have $f'(x) = (1 + x^2)/(1 - x^2)^2 > 0$ except at $x = \pm 1$. This function is increasing on each of the intervals $(-\infty, -1)$, $(-1, 1)$ and $(1, \infty)$. The graph is shown in Fig 6.4.

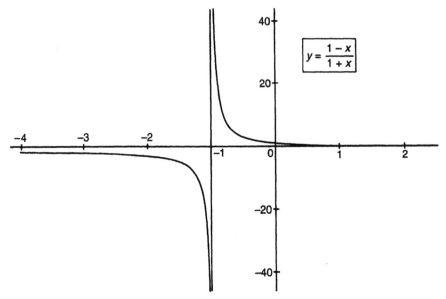

Fig 6.3 $y = (1-x)/(1+x)$.

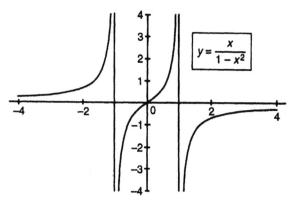

Fig 6.4 $y = x/(1-x^2)$.

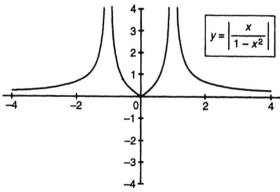

Fig 6.5 $y = |x/(1-x^2)|$.

❧ *Example 9*

Let $f(x) = |x/(1 - x^2)|$. Here again, the points $x = \pm 1$ are outside the domain of the function. Taking the absolute value changes the sign of the function for $-1 < x < 0$ and for $x > 1$, where in each case the original function defined in Example 8 was negative. In those two intervals, this changes f into a decreasing function. The graph, sketched in Fig 6.5, is increasing in each of the intervals $-\infty < x < -1$ and $0 < x < 1$, and decreasing in each of the intervals $-1 < x < 0$ and $1 < x < \infty$.

EXERCISES ON 6.2

1. Sketch the graph of the function $2x + |x|$ and verify that this function is increasing on the entire real line. By comparing the right and left derivatives at $x = 0$, verify that $2x + |x|$ is not a differentiable function.
2. Show that the functions $\tan x$ and $\tan(x/2)$ have positive derivative at each point of their respective domains. Verify the identity

$$\frac{d}{dx} \tan \frac{x}{2} = \frac{1}{1 + \cos x}$$

   For each of the functions $\tan x$, $\tan(x/2)$, find intervals on which the function is increasing. For each of the functions $\sin x$, $\sin^2 x$, find intervals on which the function is (a) increasing and (b) decreasing.
3. Find all points x at which the function $f(x) = x^2(1 - x^2)$ is (a) increasing and (b) decreasing.
4. Explain why you would expect the following 'theorem' to be true. Let a function $f(x)$ be differentiable at all x in the interval $0 \leq x \leq L$, with $f(0) = 0$ and $f'(x) < k$ for $x \in [0, L]$, where k is a constant. Then $f(L) < kL$.

   Give a formal proof of this theorem, using the Mean Value Theorem.
5. Assuming Criterion (c), given earlier, for a function to be constant, show that if two functions $f_1(x)$ and $f_2(x)$ have the same derivative at all points x of an interval $a < x < b$, then the two functions differ by a constant over the interval.

6.3 Maxima and minima

The problem of finding the maxima and minima of functions arises quite naturally and has many applications. As a simple example, suppose we have a loop of string of length l, and that we are interested in the largest area of a rectangle that can be made out of the loop. If one side of the rectangle has length x, then the other side must have length $l/2 - x$, so the area of the rectangle in that case will be $A = x(l/2 - x)$, and we want to find the largest possible value of A, as x is varied between 0 and $l/2$. A sketch of the graph $y = A(x)$ in Fig 6.6 shows that this largest value of A occurs when $x = l/4$, and is given by $A_{max} = l/4(l/2 - l/4) = l^2/16$. In fact, the rectangle having the largest possible area is a square.

We shall say that the function $A(x) = x(l/2 - x)$ has a **maximum** at $x = l/4$ and that the **maximum value** of the function is $l^2/16$. Take care to distinguish between maxima on the one hand and maximum values of functions on the other. If a

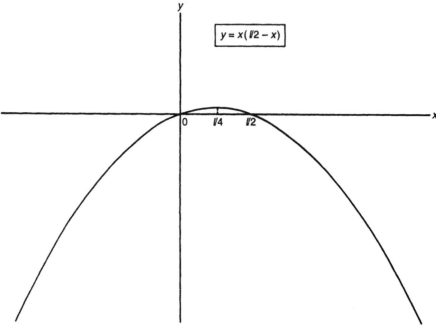

Fig 6.6 Maximum of $x(l/2 - x)$.

function f has a maximum at $x = x_0$ then the maximum value of the function will be not x_0 but $f(x_0)$.

We can also distinguish between local and global maxima. In Fig 6.7, the function $x(x^2 - 1)$ has a local maximum at $x = x_1$ and a local minimum at $x = x_2$. The value of this function at $x = x_1$ is not the largest value of $x(x^2 - 1)$ over all

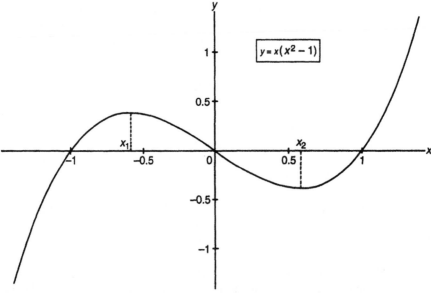

Fig 6.7 $y = x(x^2 - 1)$.

possible x values, which would be a global maximum value, but just the largest value of $x(x^2 - 1)$ if x is close to x_1, say for x in the interval $-1 \le x \le 0$. We then say that $x(x^2 - 1)$ has a local maximum at $x = x_1$ (and a local minimum at $x = x_2$). The function in this case has no global maximum, since $x(x^2 - 1)$ can be made as large as we like by taking x sufficiently large.

• Definition 2

A function f is said to have a **local maximum** at $x = x_0$ if there is an interval $a < x < b$ containing x_0 such that $f(x) \le f(x_0)$ for all x in this interval belonging to the domain of f. If $f(x) \le f(x_0)$ for **all** x in the domain of f, then f is said to have a **global maximum** at $x = x_0$. In either case, the value of $f(x_0)$ is said to be a local (respectively global) maximum value of the function f. In the case of global maxima and global maximum values, I shall often omit the word 'global' and refer simply to the maxima and maximum values of functions.

Local/global minima and minimum values are defined in a similar way. A function need have no maxima or minima at all, whether local or global. For example, this is the case for the function $f(x) = x$. Even in this case, however, there may be maxima or minima if we choose to restrict the domain. The function $f(x) = x$, with domain restricted to the interval $0 \le x \le 1$, has maximum value 1 (at $x = 1$) and minimum value 0 (at $x = 0$). The function $\sin x$, on the other hand, has infinitely many maxima, at the points $\pi/2 \pm 2n\pi$ ($n = 0, 1, 2, ...$), and infinitely many minima.

If the inequality $f(x) \le f(x_0)$ in Definition 2 can be strengthened for $x \ne x_0$ to a strict inequality $f(x) < f(x_0)$, I shall sometimes refer to **strict** (local or global) maxima, and similarly for minima. In that case, the function $\sin x$ has strict **local** maxima (since, for example, $\sin x < \sin (\pi/2)$ for all $x \in [0, \pi]$ except $x = \pi/2$), and so-called weak rather than strict **global** maxima (since, for example, there are values of x other than $\pi/2$ at which $\sin x = \sin (\pi/2)$).

Perhaps the single most important fact concerning maxima/minima of functions may be stated as follows.

Necessary criterion for maximum/minimum of a differentiable function

Suppose a function f has a maximum or minimum at $x = x_0$, which may be either local or global, and that f is differentiable at x_0. Then $f'(x_0) = 0$.

It is easy to see why this is so, using the results of Section 6.2. Take the case where x_0 is a maximum. There are only three possibilities: $f'(x_0) = 0$; $f'(x_0) > 0$; or $f'(x_0) < 0$. Consider first of all the case $f'(x_0) > 0$. Then we know that f is increasing at $x = x_0$. This means that, for some $b > x_0$, we have $f(x) > f(x_0)$ for all x in the interval $x_0 < x < b$. The value of b may be taken to be the same as the value b in Definition 2. (If the two b values differ, just choose the smaller of the two values in each case.) We now have two mutually contradictory inequalities, $f(x) > f(x_0)$ coming from $f'(x_0) > 0$, and $f(x) \le f(x_0)$ coming from the fact that there is a maximum at $x = x_0$; both inequalities hold on the interval (x_0, b). Hence our original supposition $f'(x_0) > 0$ must be incorrect. Similarly, by looking at points in the interval $a < x < x_0$, the possibility $f'(x_0) < 0$ may be ruled out. This leaves only the last remaining possibility $f'(x_0) = 0$, which is what we set out to show. A point x_0 at which $f'(x_0) = 0$ is called a **critical point** of the function f.

It is important to realize that the condition $f'(x_0) = 0$ is necessary but not sufficient for x_0 to be a maximum or minimum, if f is differentiable at x_0. By this I mean that at any maximum or minimum point x_0, provided $f'(x_0)$ exists, we can only have $f'(x_0) = 0$. On the other hand, $f'(x_0) = 0$ does not of itself guarantee that a given point x_0 is a maximum or minimum point, either local or global. If $f(x) = x^3$, then $f'(0) = 0$, whereas $x = 0$ is certainly not even locally a maximum or minimum point.

The criterion $f'(x) = 0$ does, however, give a guide in the search for maxima and minima of functions, by narrowing down the set of possible **candidates** for maxima/minima. A local maximum or minimum for a function f can occur only at points x_0 such that either

(a) f is not differentiable at x_0; or
(b) f **is** differentiable at x_0, and $f'(x_0) = 0$.

If either f is a differentiable function, in which case we need restrict our attention solely to points satisfying (b), or there are few points of non-differentiability of f, then we usually find a relatively small set of points which **may** be maxima or minima. Whether these few points are indeed maxima or minima may then be a simple matter to decide, for example by a little common sense applied to the graph of the function. In more difficult cases, the following additional criterion, which is **sufficient** to guarantee a maximum or a minimum, at least locally, will usually resolve the matter. Roughly this criterion says that a function will have a maximum at x_0 if it is increasing to the left of x_0 and decreasing to the right.

Sufficient criterion for maximum/minimum

Suppose a function f is differentiable at x_0, with $f'(x_0) = 0$. In addition, suppose that there is an interval $a < x < b$ containing x_0, such that $f'(x) > 0$ for all x in the interval $a < x < x_0$, and $f'(x) < 0$ for all x in the interval $x_0 < x < b$. Then the function f has (at least) a local maximum at x_0. On the other hand, with $f'(x_0) = 0$, if $f'(x) < 0$ for $x \in (a, x_0)$ and $f'(x) > 0$ for $x \in (x_0, b)$, then f has a local minimum at x_0.

To check the validity of this test, take the first case first. You can either use the results of Section 6.2 to see that f is increasing at each x in (a, x_0), implying that $f(x) < f(x_0)$ for all such x, or verify this directly from the Mean Value Theorem, since $[f(x_0) - f(x)]/(x_0 - x) = f'(c) > 0$ for some $c \in (a, x_0)$. Similarly, $f(x) < f(x_0)$ for all $x \in (x_0, b)$, and it follows from Definition 2 that there is a local maximum at x_0, and indeed a **strict** local maximum. (With the inequalities weakened to $f'(x) \geq 0$ and $f'(x) \leq 0$ respectively, there may possibly only be a weak local maximum.) The test for local minima is verified in a similar way.

✻ *Example 10*

A function f is given by $f(x) = x/(1 + x^2)$. Then the quotient rule gives $f'(x) = (1 - x^2)/(1 + x^2)^2$. Hence $f'(x) = 0$ at $x = \pm 1$, so that $x = +1$ and $x = -1$ are critical points. If x is greater than 1, we have $f'(x) < 0$, and if x is less than 1 (say $-1 < x < 1$) but not too much less than 1, then $f'(x) > 0$. Hence $x = 1$

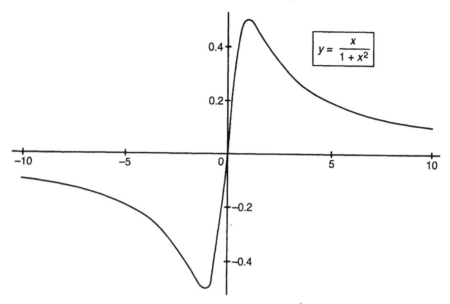

Fig 6.8 Sketching $y = x/(1 + x^2)$.

is (at least) a local maximum. A similar analysis near $x = -1$ shows that this point is a local minimum.

This information may be used to sketch the graph of $f(x)$. In sketching the graph the following observations are also useful:
(1) the domain of f is the entire real line: there are no vertical asymptotes;
(2) $f(x)$ is small and positive for large positive x, small and negative for large negative x;
(3) $f(0) = 0$, so the graph passes through the origin, which is the only point at which the graph meets the coordinate axes;
(4) $f'(0) = 1$: this gives the gradient at $x = 0$;
(5) at the maximum/minimum points already found, $f(1) = \frac{1}{2}, f(-1) = -\frac{1}{2}$.
In Fig 6.8, observations (1)–(5) are incorporated into a sketch of the graph.

Example 10 was a typical example of how to use maxima, minima and other information about functions in order to sketch their graphs. The following example adopts the same strategy in a slightly more complicated case.

Example 11

It is required to sketch the graph of the function $f(x) = (12x + 24)/(12 + x^2)$, and to find the maximum and minimum values of this function.

A simple calculation using the quotient rule shows that

$$f'(x) = \frac{-12(x^2 + 4x - 12)}{(12 + x^2)^2} = \frac{-12(x - 2)(x + 6)}{(12 + x^2)^2}$$

Hence there are critical points at $x = 2$ and at $x = -6$.

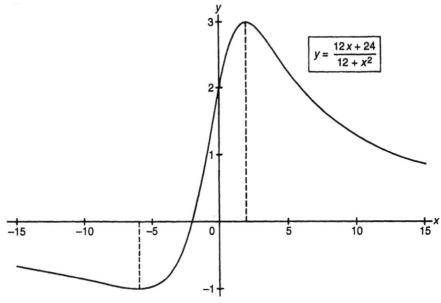

Fig 6.9   Sketching $y = (12x + 24)/(12 + x^2)$.

If x is larger than 2 we have $f'(x) < 0$ and if x is (slightly) smaller than 2 we have $f'(x) > 0$. Hence $x = 2$ is a (local or global) maximum. Similarly there is a minimum at $x = -6$. In fact a sketch of the graph (Fig 6.9) shows that these are global. Using information about large positive and negative x behaviour and intercepts with the axes, you should have little difficulty in confirming the shape of this graph, where the required maximum and minimum values of $f(x)$ are given by $f(2) = 3$ and $f(-6) = -1$, respectively.

The following example shows that the location of maxima and minima is not always simply a matter of differentiation.

⊕ *Example 12*

It is required to find the maximum and minimum values of the function $|x|/(1 + x^2)$. Now this function is not differentiable at $x = 0$, due to the $|x|$ factor in the numerator. Hence the possible candidates for maxima/minima include not only the critical points, where the derivative is zero, but $x = 0$ as well.

For $x > 0$, the derivative is

$$\frac{d}{dx}\left(\frac{x}{1 + x^2}\right) = \frac{1 - x^2}{(1 + x^2)^2}$$

and for $x < 0$ the derivative is

$$\frac{d}{dx}\left(\frac{-x}{1 + x^2}\right) = \frac{x^2 - 1}{(1 + x^2)^2}$$

Hence $x = 1$ and $x = -1$ are both critical points, and it is easy to see that these are (global) maxima, with maximum value $1/2$ in each case. On the other hand, $x = 0$ is clearly a global minimum, with minimum value zero. This minimum point would **not** be discovered if we restricted ourselves just to solutions of $f'(x) = 0$; so the function $|x|/(1 + x^2)$ has maximum value $1/2$ and minimum value 0.

The next example says much the same thing with respect to maxima/minima of functions with restricted domains.

❋ *Example 13*

What are the maximum and minimum values of the function $\psi(t) = (1 + t)/(2 + t)$ for values of t in the interval $0 \le t < \infty$?

In this case, the derivative is

$$\frac{\mathrm{d}}{\mathrm{d}t}\psi(t) = (2 + t)^{-2}$$

so there are no critical points. However, the function $\psi(t) = (1 + t)/(2 + t)$ with domain $[0, \infty)$ is **not** differentiable at $t = 0$, since this is a point on the edge of the domain. For ψ to be differentiable at $t = 0$, we would need to be in an open interval $a < t < b$ of points all in the domain of ψ. So $t = 0$ is a candidate for a maximum or minimum. Since ψ is increasing for all $t > 0$, it is clear that $t = 0$ is a minimum.

Hence, the minimum value of $(1 + t)/(2 + t)$ for $t \in [0, \infty)$ is just $1/2$. There is **no** maximum value of the function in this interval, since, for large positive values of t, $\psi(t)$ can be as close as you like to unity, without actually attaining this value. [In the language of analysis, we should say that one is the **least upper bound** for the values of $\psi(t)$ in this interval, but a least upper bound need not necessarily be attained by the function for any value of t.]

❋ *Example 14*

Let us look at an example involving a trigonometric function. Define a function ϕ by

$$\phi(x) = \frac{\sin x}{x} \quad (x \ne 0)$$
$$= 1 \quad (x = 0)$$

Since $\lim_{x \to 0} \sin x/x = 1$, the function ϕ defined in this way is a continuous function. Using the power series expansion for $\sin x$, we have

$$\phi(x) = 1 - \frac{x^2}{3!} + \frac{x^4}{5!} - \frac{x^6}{7!} \cdots$$

This series is convergent for all values of x, and agrees with $\phi(x)$ even for $x = 0$.

The series expansion is not of much use in giving a good idea of the behaviour of ϕ for large values of x. What it **is** useful for is in giving a good description of the function for values of x close to zero. In particular, we can see immediately from the series that $\phi'(x) = 0$ at $x = 0$. Hence $x = 0$ is a critical point. For small positive or negative values of x, the x term will dominate the x^3 and higher order terms in the series for $\phi'(x)$, and $\phi'(x) \approx -x/3$ will be a good approximation. Hence ϕ' is negative for small positive values of x, and positive for small negative values of x,

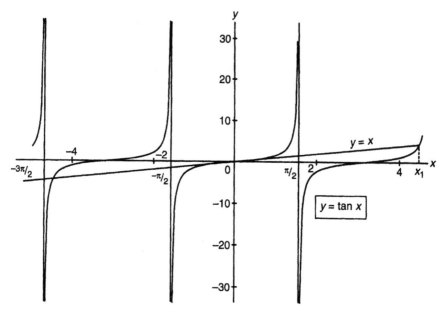

Fig 6.10 Solving tan $x = x$.

from which we may deduce that $\phi(x)$ has a maximum at $x = 0$. In fact, one may show (e.g. following the method of exercise 1 of Section 9.2) that $\sin x/x < 1$ for all $x \neq 0$, so the point $x = 0$ is a global maximum.

What other (local) maxima and minima of ϕ can we find? Using the quotient rule, we have, for $x \neq 0$, $\phi'(x) = (x \cos x - \sin x)/x^2$, so there will be a critical point at any x at which $\sin x = x \cos x$, or $\tan x = x$.

Figure 6.10 shows a sketch of part of the graph of the function $\tan x$, with the straight line $y = x$ in the same diagram. There will be a critical point at any x at which the curve $y = \tan x$ and the straight line $y = x$ intersect. One such point of intersection will occur at $x = x_1$ as indicated. From the position of the point x_1 in the diagram, note that $\cos x_1 < 0$ and that for x just to the right of x_1, $\tan x > x$. Hence, just to the right of x_1, we have $\phi'(x) = -(\cos x/x^2)(\tan x - x) > 0$, and similarly $\phi'(x) < 0$ just to the left of x_1; so $x = x_1$ is a local minimum. You should be able to see, by extending the graph, that there will be a sequence $x_1, x_2, x_3, \ldots$ of critical points, with $0 < x_1 < x_2 < x_3 \ldots$, and x_n a local minimum for n odd, and a local maximum for n even. Since ϕ is an even function ($\phi(-x) = \phi(x)$) every local maximum or minimum x_n will have a corresponding local maximum or minimum $-x_n$. The critical points x_n are the successive solutions of the equation $\tan x = x$. This equation cannot be solved exactly, and indeed cannot be solved in terms of any of the basic functions which we shall describe in this book. Nevertheless, the graphical approach which we followed in Fig 6.10 does give a good idea of the location of these critical points. It should be clear from the diagram, for example, that for large n the critical point x_n is close to, and slightly smaller than, the value $(2n + 1)\pi/2$. The value of ϕ at this critical point will then be close to

$$\sin\left(\frac{(2n + 1)\pi}{2}\right)/x_n = (-1)^n/x_n = 2(-1)^n/(2n + 1)\pi$$

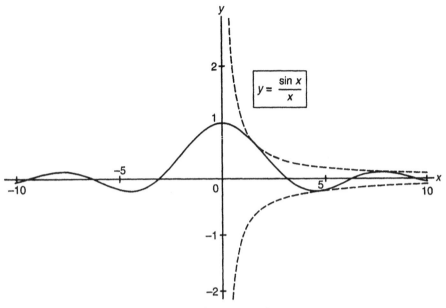

Fig 6.11 $y = \sin x/x$.

Another useful indicator in sketching the graph of ϕ comes from considering the points of intersection of the graph $y = \sin x/x$ with the curve $y = 1/x$. The two curves intersect, for $x > 0$, at $x = [(4n+1)/2]\pi$. Moreover, at each point of intersection it is easy to verify that

$$\frac{d}{dx}\left(\frac{\sin x}{x}\right) = \frac{d}{dx}\left(\frac{1}{x}\right)$$

so that the two curves have the same gradient. Now two curves having the same gradient at a point of intersection can be said to **touch** at that point. It follows that the curve $y = \sin x/x$ touches the curve $y = 1/x$ at all points $[(4n+1)/2]\pi$. In the same way, the curve touches $y = -1/x$ whenever $x = [(4n-1)/2]\pi$. I have incorporated this information in Fig 6.11 into the graph of the function ϕ.

EXERCISES ON 6.3

1. Find the maximum value of the function $x - x^3$
   (a) for all positive values of x;
   (b) for all x such that $x \geq 1$;
   (c) for all x in the interval $-2 \leq x \leq 2$.
2. Find the maximum value of the function $x\sqrt{1-x}$. Explain why this function has no minimum value.
3. Points $P_1, P_2, ..., P_n$ are located on the x-axis. If $d_i(P)$ is the distance of a general point P on the x-axis from P_i, show that the point P which minimizes $\sum_{i=1}^{n}(d_i(P))^2$ is the centroid of the set of points $\{P_i\}$. [If P_i is located at x_i and P at x, it is required to minimize the sum $\sum_{i=1}^{n}(x - x_i)^2$. The centroid of the set of points is defined to be the point having coordinate $\bar{x} = \frac{1}{n}\sum_{i=1}^{n} x_i$.]

4. With the same notation as for problem 3, express the sum $S(x) = \sum_{i=1}^{n} d_i(P)$ as a function of x. Sketch the graph of this function for some choices of the locations of the points P_i, for small values of n. Investigate how $S(x)$ changes with changes in x. Determine which points P minimize the sum $\sum_{i=1}^{n} d_i(P)$, distinguishing between the cases of n even and n odd.

6.4 More about second derivatives

Traditionally, the second derivative has often been thought of as playing an important role in the characterization of maxima and minima of functions. Often, the importance of evaluating second derivatives at critical points to test for maxima and minima is as a theoretical rather than a practical tool. It is usually simpler, in the case of differentiable functions, to work with the the first derivative and use the criteria which have already been described in Section 6.3, which are based on first rather than second derivatives. Despite all this, the second derivative is rather special, particularly in applications to mechanics, and deserves further attention.

In Section 6.2, we saw that a function f is increasing at $x = x_0$ if $f'(x_0) > 0$, and decreasing if $f'(x_0) < 0$. What does the sign of the **second** derivative tell us about the behaviour of a function? Now the second derivative $f''(x)$, also written as d^2f/dx^2, is just the derivative of the first derivative; $d^2f/dx^2 = (d/dx)(df/dx)$; so $f''(x_0) > 0$ tells us that $f'(x)$ is increasing at x_0, and similarly if $f''(x_0) < 0$ then $f'(x)$ is decreasing at x_0. In terms of the graph $y = f(x)$, we can say that the slope increases at points where $f''(x) > 0$, and decreases at points at which $f''(x) < 0$ (see Fig 6.12).

As a special case of this result, consider a critical point x_0, at which $f'(x_0) = 0$. If $f''(x_0) > 0$, it follows that there is an interval $a < x < b$ containing x_0 such that $f'(x) > 0$ for $x > x_0$ and $f'(x) < 0$ for $x < x_0$. Thus our criterion for a function to have a (local) minimum is satisfied. The minimum occurs at $x = x_0$. By the same argument, if $f'(x_0) = 0$ and $f''(x_0) < 0$ then $x = x_0$ will be a local maximum.

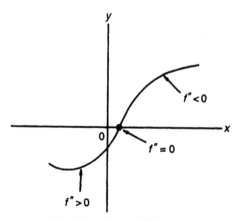

Fig 6.12 Sign of $f''(x)$.

It is clearly easier to evaluate a first derivative than a second derivative, and sometimes the so-called second derivative test for maxima/minima is difficult to implement in practice.

⚛ *Example 15*

A function f is given by $f(x) = (1 - x)/(1 + x)^2$. By the quotient rule, for $x \neq -1$ we have $f'(x) = (x - 3)/(1 + x)^3$. Hence there is a critical point at $x = 3$. At this critical point, the second derivative is $[(1 + x)^3 - 3(x - 3)(1 + x)^2]/(1 + x)^6 = 1/4^3 > 0$.

Hence $x = 3$ is a minimum; notice, however, that this is also clear from the observation that $f'(x) > 0$ for $x > 3$ and $f'(x) < 0$ for $-1 < x < 3$.

The graph $y = f(x)$ is indicated in Fig 6.13. In fact, $x = 3$ is a global minimum, and the minimum value of $f(x)$ at this point is $-1/8$. Observe also that $f''(x) = 2(5 - x)/(1 + x)^4 > 0$ for $x < 5$, and that $f''(x) < 0$ for $x > 5$. In the diagram, the graph should be 'concave upwards' for $x < 5$, and 'concave downwards' for $x > 5$.

Earlier on, I referred to the role played by velocity in mechanics. For a particle of mass m, moving along the x-axis, with coordinate given as a function of time by $x = x(t)$, the velocity at time t is dx/dt, and the particle has momentum $m(dx/dt)$. The acceleration of the particle is then defined to be the rate of change of velocity, and at time t is given by $(d/dt)(dx/dt) = d^2x/dt^2$; and the total force acting on the particle, according to Newton's second law of motion, is given by the rate of change of momentum. As long as the mass m does not change with time, the force is then given by

$$F = \frac{d}{dt}\left(m\frac{dx}{dt}\right) = m\frac{d^2x}{dt^2}$$

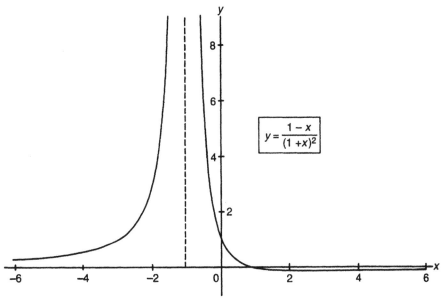

Fig 6.13 $y = (1 - x)/(1 + x)^2$.

The fact that Newtonian mechanics is firmly based on the concept of force implies that the equations of motion, that is those equations which describe the dynamics of a classical Newtonian system, can be formulated in terms of second derivatives. We are led, in fact, to second order differential equations, about which I shall have more to say in Chapters 10 and 11.

For a single particle moving freely, that is not subject to a force, the acceleration will be zero. In that case, $(d/dt)(dx/dt) = 0$, from which it follows that the velocity dx/dt remains constant. This is essentially Newton's law of inertia, or first law of motion, which states that a freely moving particle will move in a straight line with constant velocity. Any change in velocity can result only from an applied force, and Newton's second law then gives a direct relationship between the applied force and the rate of change of velocity.

Of course most dynamical systems show a greater degree of complexity than is the case for a single particle moving in a straight line. In general, sufficiently many coordinates must be introduced in order to describe the motion of all of the particles of the system, and these may move in three dimensions. Nevertheless, the basic equations will still involve second derivatives of these coordinates. Why this is so is an interesting question to which I shall return later, in Chapter 10.

EXERCISES ON 6.4

1. The function f has a critical point $x = x_0$, so that $f'(x_0) = 0$. It is also known that $f''(x_0) = 0$, and that $f'''(x_0) = 1$. Is $x = x_0$ a maximum or a minimum? What can be said about the behaviour of the function at $x = x_0$? Give an example of a function f such that $f(0) = 1, f'(0) = 0, f''(0) = 0$, and $f'''(0) = 1$.
2. The function g has a critical point $x = x_0$, so that $g'(x_0) = 0$. It is also known that $g''(x_0) = 0$, $g'''(x_0) = 0$, and that $g''''(x_0) = 1$. Show that x_0 is a local minimum for g, and give an example of such a function g.
3. A particle moves along the x-axis with constant acceleration g, so that $d^2x/dt^2 = g$, where $x(t)$ is the coordinate of the particle at time t. By considering the derivatives

$$\frac{d}{dt}\left(\frac{dx}{dt} - gt\right) \text{ and } \frac{d}{dt}\left(x - \frac{1}{2}gt^2 - ut\right)$$

show that

$$x(t) = \frac{1}{2}gt^2 + ut + a$$

where u and a are the respective values of dx/dt and x when $t = 0$.

Summary

We have seen many ways in this chapter in which derivatives can help us to understand the behaviour of functions. In many cases, and this applies to most of the standard functions which are the most frequently used, a knowledge of first and higher order derivatives at just a single point will enable us to write down a power series expansion. This is the so-called Taylor series for a function f. The Taylor series about $x = 0$ is a series of powers of x, and more generally we can write down a series about $x = x_0$, in powers of $(x - x_0)$. Such series are often of use in studying functions, and in Chapter 7 we shall investigate the exponential function starting from a power series.

The first derivative df/dx tells us about the rate of change of the values of a function f, and determines whether f is increasing or decreasing at a point, or in an interval. All this provides useful information in sketching the graph, looking at asymptotic behaviour, and so on. In the case of critical points, where $df/dx = 0$, the local variation of the derivative should allow us to decide whether we are dealing with a maximum, minimum, or neither of these. In principle, in suitable cases, the second or higher derivatives can also help in analysing critical points, though I would not wish to stress the use of these where other, more simple and direct, methods are available.

FURTHER EXERCISES

1. Write down the Taylor series about $x = 1$ for the function x^4.
2. Obtain the Taylor series (a) about $x = 0$, and (b) about $x = \pi/2$, for the function $\cos x$.
3. If $f(x) = (1 + x)^{-1/2}$, show that the nth derivative of f at $x = 0$ is given by

$$f^{(n)}(0) = \frac{(-1)^n.1.3...(2n - 1)}{2^n}$$

   Hence write down the Taylor series about $x = 0$ for the function $f(x)$, and use the ratio test to show that the series has radius of convergence $R = 1$. Explain why a radius of convergence larger than unity would not have been expected.
4. In Section 6.2 it is shown that any function having zero derivative on an interval is a constant function. With the assumption and notation of problem 2 in the exercises at the end of Chapter 5, and assuming that the functions c and s have values at $x = 0$ given by $c(0) = 1$ and $s(0) = 0$ respectively, show that $c^2 + s^2 = 1$ for all x. Hence show that

$$\frac{d}{dx}\left(\frac{s}{c}\right) = \frac{1}{c^2}, \frac{d}{dx}\left(\frac{c}{s}\right) = -\frac{1}{s^2}$$

5. At which values of x is the function $2x - x^3$ increasing, and at which values of x is the function decreasing? Find the maximum value of $2x - x^3$, for all positive values of x.
6. Sketch the graph of the function $1/(x^2 - x - 2)$, and show that this function has a local maximum at $x = \frac{1}{2}$. Find the maximum value of this function (a) for values of x in the interval $-1 < x < 1$ and (b) for values of x in the interval $1 \leq x < 2$.
7. Verify that the function $g(x) = (2x - 3)/(2 - x^2)$ has critical points at $x = 1$ and at $x = 2$. Determine the nature of these critical points, i.e. whether they are local maxima or local minima. Sketch the graph of the function $g(x)$.
8. Find the maxima and minima of the function $|x - x^3|$, and determine whether they are local or global.
9. The following method can be used to find approximate roots to equations $f(x) = 0$, and is called the Newton–Raphson method.

Let $f(x)$ be a function having derivatives of first and second order; let $x = x_0$ be a root of the equation $f(x) = 0$. An approximate value of the root, $x = x_1$, is given, so x_1 is close to x_0, and it is required to replace x_1 by a better approximation x_2, which is closer to x_0.

Define a function g by $g(x) = x - [f(x)/f'(x)]$. Assuming that $f'(x_0) \neq 0$, show that (a) $g(x_0) = x_0$, (b) $g'(x_0) = 0$ and (c) $g''(x_0) = f''(x_0)/f'(x_0)$. Hence use (a) and (b) to show that

$$\lim_{x_1 \to x_0} \frac{g(x_1) - x_0}{x_1 - x_0} = 0$$

For x_1 close to x_0, explain why $x_2 = g(x_1)$ is a better approximation to x_0 than x_1. If a sequence of approximations to x_0 is defined iteratively by $x_{n+1} = g(x_n)$, show that $x_n \to x_0$ as $n \to \infty$. Use this method to obtain a root of the equation $x^3 - x - 1 = 0$, between $x = 1$ and $x = 2$, to six decimal places.

7 • Some Special Functions

It is difficult to imagine what the mathematical world would be like without the all pervasive exponential function, and its close cousin and inverse function the logarithm. If these functions were no longer available, a considerable body of mathematics, at all levels, would either disappear or at most struggle on with difficulty. The exponential function, for example, plays a central role in the precise definition of non-integral powers of x, in the evaluation of various limits, and in the solution of all manner of differential equations. Without the logarithmic function, we could no longer claim to be able to integrate all powers of x. Many of the applications of mathematics involve these functions in a crucial way.

What is it that makes these functions so special? I shall try to answer this question, starting with the exponential function defined as a power series. We shall find that everything that can be said about the exponential and logarithm starts from this very simple beginning. After that, we shall look at some other examples of rather special functions, which are closely linked to the exponential function, and see how they too have important contributions to make to our subject.

7.1 The exponential function

The exponential function $\exp(x)$ is defined for all $x \in \mathbb{R}$ by the power series

$$\exp(x) = 1 + x + \frac{x^2}{2!} + \frac{x^3}{3!} + \frac{x^4}{4!} + \ldots \tag{7.1}$$

We have already seen, by using the ratio test in Section 5.3, that this series is convergent for all values of x. The series has infinite radius of convergence, which implies in particular that there is no difficulty in differentiating term by term to determine the derivative $\mathrm{d}(\exp(x))/\mathrm{d}x$. The derivative is given by the differentiated power series

$$1 + \frac{2x}{2!} + \frac{3x^2}{3!} + \frac{4x^3}{4!} + \ldots = 1 + x + \frac{x^2}{2!} + \frac{x^3}{3!} + \frac{x^4}{4!} + \ldots \tag{7.2}$$

Either you have carried out this derivative before, and know what the answer is, in which case you have probably by now got used to the result — or if not, this may be a minor miracle to you. The fact of the matter is that the exponential function differentiates to give itself; that is, we find that

$$\frac{\mathrm{d}}{\mathrm{d}x}\exp(x) = \exp(x)$$

Thus the operation of differentiation, which as we have seen in Chapter 5 is a linear operation, and in Chapter 4 an operation which tells us what we need to know about rates of change, and gradients, and so much more about functions in

Chapter 6, is an operation which leaves the exponential function completely unaltered.

Such a property of exp(x) may be expected to be important, and perhaps to have some remarkable consequences. Is this property of exp(x) unique to the exponential function, or might there be other functions f which also satisfy $df/dx = f$ for all values of x? First of all, we should recognize that any constant multiple of exp(x) will also have the property $df/dx = f$, since if $f = A\exp(x)$ with A a constant then

$$\frac{df}{dx} = \frac{d}{dx}A\exp(x) = A\frac{d}{dx}\exp(x) = A\exp(x) = f$$

as required. In fact, any function f satisfying $df/dx = f$ can **only** be a constant multiple of exp(x). To see why this is so, we need to look for a moment at the function exp($-x$).

Now, if $y = \exp(t)$ with $t = -x$, then the chain rule tells us that $dy/dx = (dy/dt).(dt/dx) = (\exp(t)) \times -1 = -\exp(t) = -\exp(-x)$. Hence the function exp($-x$) satisfies the equation

$$\frac{d}{dx}\exp(-x) = -\exp(-x)$$

A simple consequence of this result, on using the product rule, is that

$$\frac{d}{dx}\{\exp(-x) \times \exp(x)\} = \exp(-x)\frac{d}{dx}\exp(x) + \exp(x)\frac{d}{dx}\exp(-x)$$
$$= \{\exp(-x) \times \exp(x)\} + \{\exp(x) \times -\exp(-x)\}$$
$$= 0$$

Hence the function exp($-x$) × exp(x) has zero derivative for all values of x, and is therefore a constant function. Substituting $x = 0$ into the power series for the exponential function gives exp(0) = 1. Hence also the constant function exp($-x$) × exp(x) has the value 1 at $x = 0$, and it follows that exp($-x$) × exp(x) = 1 for **all** x. We are now ready to find **all** functions f which satisfy $df/dx = f$.

Uniqueness of the exponential function

Let f be any function such that $df/dx = f$ for all $x \in \mathbb{R}$. Then f is a constant multiple of the exponential function. If f satisfies the further condition $f(0) = 1$, then f is in fact the exponential function.

PROOF

Suppose f is any function for which $df/dx = f$ for all $x \in \mathbb{R}$. We have to show that $f(x) = A\exp(x)$ for some constant A. Multiplying both sides by exp($-x$) and using the identity exp($-x$) × exp(x) = 1, this amounts to showing that exp($-x$) × $f(x)$ is a constant function. There is an easy way to do this — it is only necessary to verify that

$$\frac{d}{dx}\{\exp(-x)f(x)\} = 0$$

for all $x \in \mathbb{R}$. The fact that the derivative is always zero will then imply $\exp(-x)f(x) = \text{constant} = A$, say, which is the result we want.

Well, by the product rule,

$$\frac{d}{dx}\{\exp(-x)f(x)\} = \exp(-x)\frac{d}{dx}f(x) + f(x)\frac{d}{dx}\exp(-x)$$

$$= \exp(-x)\frac{d}{dx}f(x) + f(x) \times -\exp(-x)$$

$$= \exp(-x)(\frac{d}{dx}f(x) - f(x)) = 0$$

as required, since $df/dx = f$. Hence we can indeed assert that $f(x) = A\exp(x)$, and the first part of the result is proved.

Knowing that a function satisfying $df/dx = f$ must be of the form $f(x) = A\exp(x)$, it follows straight away that if f has the further property $f(0) = 1$, then $f(0) = A\exp(0) = A = 1$, so that $f(x) = \exp(x)$ is the only possibility. The exponential function, then, is uniquely determined by the two properties (i) $df/dx = f$ for all $x \in \mathbb{R}$, and (ii) $f(0) = 1$.

What kind of a function is $\exp(x)$? Notice first of all, from the power series expansion in equation (7.1), that $\exp(x)$ tends to infinity as x tends to infinity very rapidly indeed, in fact more rapidly than any power of x. For example, given any positive integer N the series contains a term $x^{N+1}/(N+1)!$. This tells us that, even taking no account of all the other terms of the series, which must, however, be positive for x positive, we have $\exp(x) > x^{N+1}/(N+1)!$ for $x > 0$. Hence $\exp(x)/x^N > x/(N+1)!$, which diverges in the limit as $x \to \infty$; so $\exp(x)$ is much larger than x^N, for sufficiently large x.

Although not immediately evident from the power series, it is also true that $\exp(x) > 0$ for all x. For $x \geq 0$ this follows from the power series for $\exp(x)$, and for $x \leq 0$ it comes from the identity $\exp(-x).\exp(x) = 1$. Moreover, $\lim_{x\to-\infty}\exp(x) = \lim_{x\to\infty}\exp(-x) = \lim_{x\to\infty}1/\exp(x) = 0$. Since $\exp(x)$ tends to infinity as x tends to infinity more rapidly than any positive power of x, the convergence of $\exp(x)$ to zero as $x \to -\infty$ is more rapid than any **negative** power of x, in the sense that $\lim_{x\to-\infty}|x|^N\exp(x) = 0$ for any $N > 0$.

We have enough information now to have a good idea of the general shape of the graph of the exponential function. Notice too that the property $df/dx = f$, with $f > 0$, guarantees not only that the gradient of the exponential function is positive, but on repeated differentiation we find $d^nf/dx^n = f$ as well, so that all higher order derivatives are also positive. Figure 7.1 shows the graph of the exponential function. Just as the exponential function is a rather special function in mathematics, the value of this function at $x = 1$ is a rather special number. We shall denote this number by the symbol e. Thus $\exp(1) = e$.

The series for $\exp(x)$ is very rapidly convergent if x is not taken too large, and the sum of the first ten or so terms at $x = 1$ will give a very good estimate of the value of e. In fact, $e = 1 + 1 + \frac{1}{2!} + \frac{1}{3!} + \frac{1}{4!} + \ldots = 2.7182818\ldots$. The decimal expansion of e is not a recurring decimal, and there is no simple pattern to be found in the later decimals of the expansion. Mathematicians have done a great deal of work to try to discover the properties of e. It is known that e is both irrational (that is it cannot be expressed as the ratio of two integers) and even transcendental (that is,

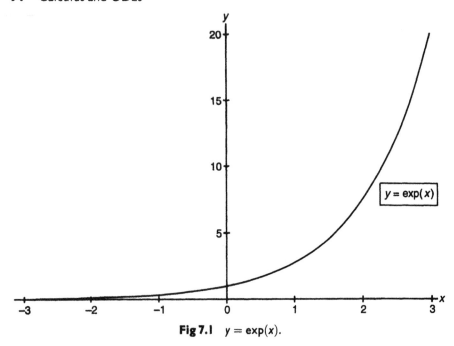

Fig 7.1 $y = \exp(x)$.

there is no polynomial equation with integer coefficients of which e is a solution). In the end, though much is known, a great deal will remain elusive about this mysterious number.

EXERCISES ON 7.1

1. Use the power series expansion for the exponential function to confirm the above value of the constant e to seven decimal places. Use the series to calculate $\exp(-1)$ to a similar degree of accuracy, and verify that $\exp(-1) = 1/e$ to this approximation.
2. A function f is known to satisfy the equation $df/dx = 2f$ for all $x \in \mathbb{R}$. Show that f is a constant multiple of $\exp(2x)$, and that if f satisfies the additional property $f(0) = 1$, then $f(x) = \exp(2x)$. Use the product rule for differentiation to verify that $f(x) = (\exp(x))^2$ satisfies the equation $df/dx = 2f$, and hence prove the identity $(\exp(x))^2 = \exp(2x)$. Use this identity to express $\exp(2)$, $\exp(4)$, $\exp(1/2)$ and $\exp(-1/2)$ in terms of the constant e.

7.2 Exponential and logarithm

A look at the graph of the exponential function in Fig 7.1 should convince you that this is an injective function, and hence that it is possible to define an inverse function (see Section 3.2 for the definition of an injective function, and for the relevance of this concept to inverse functions).

The inverse function to $\exp(x)$ is defined to be the **natural logarithm**, or just **logarithm** for short, and will be written $\ln(x)$. From the definition of inverse functions, $y = \ln(x)$ means the same as $x = \exp(y)$. Since $\exp(y)$ is always positive, $\ln(x)$ is defined only for positive values of x. Just as the exponential function has domain $\mathbb{R}$ and range $\mathbb{R}_+(= (0, \infty))$, so the logarithmic function has domain $\mathbb{R}_+$ and range $\mathbb{R}$. The graph of $\ln(x)$ is shown in Fig 7.2, and follows directly from the sketch of $y = \exp(x)$ in Fig 7.1, on exchanging the roles of x and y.

Certain properties of $\ln(x)$ follow straight away from corresponding properties of $\exp(x)$. We may deduce, for example, that $\ln(1) = 0$ and $\ln(e) = 1$. As x approaches zero from above, $\ln(x)$ tends to $-\infty$. As x tends to $+\infty$, $\ln(x)$ diverges very slowly indeed.

At this stage, probably the most important results to remember for $\exp(x)$ and $\ln(x)$ are those which link the two functions on account of each function being the inverse of the other. This may be summarized in the two basic equations

$$\left.\begin{array}{l} \ln(\exp x) = x, \text{for all } x \in \mathbb{R} \\ \exp(\ln x) = x, \text{for all } x > 0 \end{array}\right\} \tag{7.3}$$

It is, I hope, a straightforward exercise in the differentiation of inverse functions, following the ideas and methods of Section 5.4, to determine the derivative of $\ln(x)$.

If $y = \ln(x)$, then $x = \exp(y)$, so that $\mathrm{d}x/\mathrm{d}y = \exp(y)$. Hence $\mathrm{d}y/\mathrm{d}x = 1/\exp(y)$, which on expressing the right-hand side as a function of x leads to the result

$$\frac{\mathrm{d}}{\mathrm{d}x}\ln(x) = \frac{1}{x} \quad (x > 0) \tag{7.4a}$$

Equation (7.4a) fills a gap in the table of derivatives. If we consider the sequence of functions $...x^3, x^2, x, 1, 1/x, 1/x^2, ...$ (in which each successive function is obtained

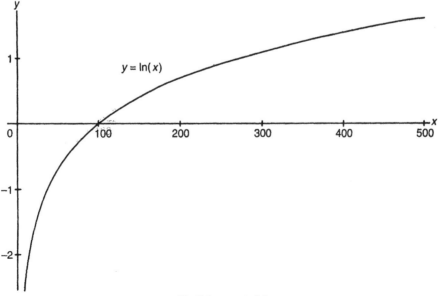

Fig 7.2 $y = \ln(x)$.

from the previous function by dividing by x), together with the corresponding sequence of derivatives $...3x^2, 2x, 1, 0, -1/x^2, -2/x^3, ...$, you will see that there is no function in the sequence of derivatives containing the power $1/x$. Equation (7.4a) tells us why this is so — to arrive at an inverse power of x we need to differentiate a logarithmic function, not a power. This curious anomaly has profound consequences, as we shall see, in the theory of integration.

Equation (7.4a) does not make sense for $x < 0$, where $\ln(x)$ is not even defined. There is, however, a corresponding result for negative values of x, which comes from evaluating the derivative of $\ln(-x)$. Applying the chain rule, with $y = \ln(t)$, $t = -x$, we find $dy/dt = 1/t$, $dt/dx = -1$, so that $dy/dx = (dy/dt).(dt/dx) = -1/t = 1/x$. Hence

$$\frac{d}{dx}\ln(-x) = \frac{1}{x} \quad (x < 0) \tag{7.4b}$$

This is the same derivative as before, this time for negative values of x. It is often convenient to combine equations (7.4a) and (7.4b) into a single equation, noting that $|x|$ is just x for $x > 0$ and $-x$ for $x < 0$. The resulting single equation now becomes

$$\frac{d}{dx}\ln(|x|) = \frac{1}{x} \quad (x \neq 0) \tag{7.5}$$

Equation (7.5) now holds for all $x \in \mathbb{R}$ except at $x = 0$. Since the function $\ln(|x|)$ has a derivative at each point of its domain we have shown that $\ln(|x|)$ is a differentiable function.

I shall now use almost the same argument that led to the derivative of the natural logarithm to prove a pair of remarkable identities, which take us to the heart of what makes the exponential and logarithm so special.

Two identities for logarithm and exponential

The logarithmic and exponential functions satisfy the respective identities

$$\ln(ab) = \ln(a) + \ln(b) \quad (a, b > 0) \tag{7.6}$$
$$\exp(A + B) = \exp A \times \exp B \quad (A, B \in \mathbb{R}) \tag{7.7}$$

PROOF
The key to the proof of these two identities is the evaluation of the derivative $d(\ln(ax))/dx$, for $x > 0$, where a is a positive constant. This is easily done using the chain rule, and $y = \ln(t)$, $t = ax$ gives $dy/dt = 1/t$, $dt/dx = a$, so that $dy/dx = (dy/dt).(dt/dx) = a/t = 1/x$. Hence

$$\frac{d}{dx}(\ln(ax)) = \frac{1}{x} \text{ for } x > 0$$

where a is any positive constant. However, we know that

$$\frac{d}{dx}(\ln(x)) = \frac{1}{x} \text{ for } x > 0$$

in fact this is the same result, for the special case $a = 1$.

Subtracting now gives

$$\frac{d}{dx}\{\ln(ax) - \ln(x)\} = \frac{1}{x} - \frac{1}{x} = 0$$

for all $x > 0$. Hence $\ln(ax) - \ln(x)$ is a constant function on the interval $0 < x < \infty$. Moreover, at $x = 1$ this constant function has the value $\ln(a) - \ln(1) = \ln(a)$. It follows that $\ln(ax) - \ln(x) = \ln(a)$, and the identity (7.6) comes directly from substituting $x = b$.

To derive the second of the two identities, substitute $a = \exp A$ and $b = \exp B$ in equation (7.6), noting that both a and b defined in this way are necessarily positive. Then $\ln(\exp A \times \exp B) = \ln(\exp A) + \ln(\exp B) = (A + B)$ since ln is the inverse function to exp. Hence $\ln(x) = y$, where $x = \exp A \times \exp B$ and $y = A + B$. Again using the inverse properties of the logarithmic and exponential functions to rewrite the equation $x = \exp(y)$, the identity (7.7) is now proved.

The two identities (7.6) and (7.7) show that in some sense the logarithm and exponential are able to interchange the roles of addition and multiplication of real numbers. Thus the logarithm of a **product** is the **sum** of the logarithms, and the exponential of a **sum** is the **product** of the exponentials. This will have an important bearing on how to define and evaluate powers, as we shall see in the following section.

EXERCISES ON 7.2

1. Use the chain rule to show that

$$\frac{d}{dx}\ln(\sqrt{x}) = \frac{1}{2x}, \text{ for } x > 0$$

Hence show that

$$\ln(\sqrt{x}) = \frac{1}{2}\ln(x)$$

How could this result be deduced directly from equation (7.6)?

2. Show that

$$\frac{d}{dx}\ln(1 + x) = \frac{1}{1 + x}$$

(a) by assuming the derivative of $\ln(x)$, and using the chain rule, and (b) by differentiating the equation $1 + x = \exp y$. Determine the nth order derivative of $\ln(1 + x)$ at $x = 0$, and hence obtain the Taylor expansion

$$\ln(1 + x) = x - \frac{x^2}{2} + \frac{x^3}{3} - \frac{x^4}{4} \ldots$$

What is the radius of convergence of this series?

3. Use the chain rule to show that

$$\frac{d}{dx}\exp(x + c) = \exp(x + c)$$

Explain how this result can be used to provide an alternative proof of the identity $\exp(x + c) = \exp x \times \exp c$.

4. Show how equations (7.6) and (7.7) may be used to obtain the identities
$\ln(a/b) = \ln(a) - \ln(b)$; $\exp(A - B) = \exp A / \exp B$; $-\ln(a) = \ln(1/a)$;
$\exp(-A) = 1/\exp A$.

7.3 Powers and their inverses

The identities for exponentials and logarithms lead naturally to a new way of looking at powers of x. To see this, consider how you would evaluate the logarithm of a power of x. Starting from small integer powers, this is quite straightforward. For example, $\ln(x^2) = \ln(x \times x) = \ln(x) + \ln(x) = 2\ln(x)$. $\ln(x^3) = \ln(x \times x^2) = \ln(x) + \ln(x^2) = \ln(x) + 2\ln(x) = 3\ln(x)$. The pattern continues in an obvious way, and we arrive at the result $\ln(x^n) = n\ln(x)$, for any positive integer n and all $x > 0$.

For negative integer powers, we can use the identity $\ln(1/a) = -\ln(a)$, derived in problem 4 of Exercises on 7.2. This gives $\ln(x^{-n}) = \ln(1/x^n) = -\ln(x^n) = -n\ln(x)$, for any positive integer n and all $x > 0$.

In all these cases, then, we have the common result $\ln(x^c) = c\ln(x)$, which holds for all $x > 0$, where c can be any integer, positive or negative. The identity can also be extended to any **rational** value of c; the case $c = 1/2$, for example, follows from problem 1 of Exercises on 7.2.

Let us, then, invert the equation $\ln(x^c) = c\ln(x)$ to give

$$x^c = \exp(c\ln(x)) \quad (x > 0) \tag{7.8}$$

and use this as the defining equation for powers of a positive number x. This equation certainly agrees with what we know already of integer and rational powers of x, and it seems a small further step to use equation (7.8) for **any** real value of c. The fact that equation (7.8) does not work for negative values of x is no great drawback, since there are relatively very few cases (say x^{-1} and $x^{1/3}$ for example, but not $x^{1/2}$ or $x^{1/4}$) for which x^c makes sense in any case with $x < 0$.

Starting from equation (7.8), all the standard identities for powers can be verified, and follow from properties of the logarithm and exponential that we have already seen. As a simple example, we have

$$\begin{aligned} x^a \times x^b &= \exp(a\ln(x)) \times \exp(b\ln(x)) \\ &= \exp((a + b)\ln(x)) \\ &= x^{(a+b)} \end{aligned}$$

There is another interesting and fundamental consequence of our formula for powers. It is that the exponential function is itself a power. Replacing c by x and x by e in equation (7.8), we arrive at the result $e^x = \exp(x\ln(e)) = \exp(x)$, since $\ln(e) = 1$. Thus the exponential function is none other than the number e, raised to the power x. No wonder, then, that $\exp(x)$ diverges so rapidly in the limit as x tends to infinity — as x increases, the number e is raised to an ever larger power; and the behaviour for x large and negative is also well understood when we see that

$\exp(x) = 1/\exp(-x) = 1/e^{-x}$, which tends to zero very rapidly in the limit $x \to -\infty$.

Just as the natural logarithm $\ln(x)$ is the inverse function of e^x, we define the inverse function of a^x, for any $a > 0$, to be the logarithm function **to the base** a, and write this function $\log_a x$. Then $y = \log_a x$ means the same as $x = a^y$. In other words, the logarithm of x, to the base a, is the power to which a must be raised, in order to give the value x. In the special case $a = e$, we have $\log_e x = \ln(x)$, and the natural logarithm may be identified with the logarithm to the base e. Apart from the natural logarithm, the other logarithm which is most commonly seen is $\log_{10} x$, the logarithm to the base 10.

Let me return to a question which I asked at the beginning of this chapter. Namely, what makes $\ln(x)$ so special among functions? I think that we have seen part of the answer already, in the remarkable properties which this function and its inverse, the exponential, possess. Now it is possible to pose the same question in a rather more special way. Of all the logarithmic functions, that is logarithms to various bases, what makes the natural logarithm, the logarithm to the base e, stand out? And how are logarithms to different bases related to each other?

To answer these questions, we have to express logarithms to the base a in terms of logarithms to the base e, so let $y = \log_a x$, implying $x = a^y$. Now take logarithms of both sides of this equation, **to the base** e. This gives $\ln(x) = y \ln(a)$, implying that $y = \ln(x)/\ln(a)$. This, then, is the link between logarithms to different bases. We have $\log_a x = \log_e(x)/\log_e(a)$; logarithms to different bases are proportional to one another.

Since

$$\frac{d}{dx}\log_e(x) = \frac{d}{dx}\ln(x) = \frac{1}{x} \text{ (for } x > 0)$$

we can also use this link between different bases to determine the derivative of $\log_a x$. In fact, we have

$$\frac{d}{dx}\log_a x = \frac{1}{x \ln(a)}$$

Only in the single case $a = e$, for which $\ln(e) = 1$, do we have

$$\frac{d}{dx}\log_e x = \frac{1}{x}$$

In all other cases there is a constant of proportionality. This property alone singles out $\ln(x)$ as a rather special function; it is aptly named the natural logarithm!

EXERCISES ON 7.3

1. Use the definition of powers given in equation (7.8), together with properties of the exponential and logarithmic function, to verify the results $(x^a)^b = x^{ab}$, $x^0 = 1$, and $x^{1/n} = \sqrt[n]{x}$, where n is any positive integer.
2. Explain how you would define x^c $(x > 0)$ without using logarithms or exponentials (a) if c is a (positive or negative) integer, and (b) if c is a rational number. Why does your definition not allow you to define $x^{\sqrt{2}}$? Use the

definition of powers given in equation (7.8) to show that x^c depends continuously on c, for fixed x. How does this help in defining the value of $x^{\sqrt{2}}$?

3. Evaluate $\log_2 16$, $\log_{10}(0.001)$ and $\log_3(1/3)$, and show that $\log_3 5$ is an irrational number.

4. Discover what you can about the history of the invention of logarithms. Give a short description of (a) Napier's bones, and (b) the sliderule.

7.4 Sines, cosines, hyperbolic functions — and complex numbers

The hyperbolic functions are a special kind of function which can be expressed very simply in terms of exponentials, and which, moreover, are very closely related to trigonometric functions. The hyperbolic function which relates most closely to the sine is called the sinh x (pronounced 'synch x'). Similarly, there is a hyperbolic cosine, written cosh x (and pronounced just as it looks) and a hyperbolic tangent, defined as you would expect by tanh $x = \sinh x/\cosh x$ (and pronounced 'tanch x'). The definitions of sinh x and cosh x are as follows: sinh $x = (e^x - e^{-x})/2$, cosh $x = (e^x + e^{-x})/2$, so that correspondingly we have tanh $x = (e^x - e^{-x})/(e^x + e^{-x})$. The graphs of these functions are sketched in Fig 7.3.

Although the graphs of sinh, cosh and tanh hardly resemble those of their trigonometric counterparts, the hyperbolic functions satisfy a number of identities which are very similar to corresponding results for sine, cosine and tangent. In most cases the only differences between the two groups of identities lie in one or two changes in sign here and there. As examples of this, we have

$$\frac{d}{dx}\sinh x = \frac{d}{dx}\left(\frac{e^x - e^{-x}}{2}\right) = \frac{e^x + e^{-x}}{2} = \cosh x$$

which corresponds faithfully to the trigonometric identity

$$\frac{d}{dx}\sin x = \cos x$$

however, the derivative of the hyperbolic cosine function is not $-\sinh x$ as we might expect but

$$\frac{d}{dx}\cosh x = \frac{d}{dx}\left(\frac{e^x + e^{-x}}{2}\right) = \frac{e^x - e^{-x}}{2} = \sinh x$$

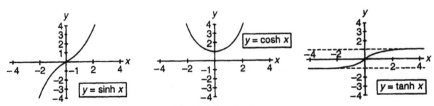

Fig 7.3 Graphs of hyperbolic functions.

Perhaps the most fundamental identity of all, relating sine and cosine, is $\cos^2 x + \sin^2 x = 1$. Here the corresponding hyperbolic identity again involves a change of sign, and takes the form $\cosh^2 x - \sinh^2 x = 1$. This identity may be verified directly from the definitions of $\cosh x$ and $\sinh x$, using standard properties of the exponential function. Since $\cosh x + \sinh x = ((e^x + e^{-x})/2) + ((e^x - e^{-x})/2) = e^x$, and similarly $\cosh x - \sinh x = e^{-x}$, we have

$$\cosh^2 x - \sinh^2 x = (\cosh x + \sinh x)(\cosh x - \sinh x) = e^x . e^{-x} = 1$$

as required.

A comparison between the two identities $\cos^2 x + \sin^2 x = 1$ and $\cosh^2 x - \sinh^2 x = 1$ also helps to explain the reason for the description 'hyperbolic', if you recall that $X^2 + Y^2 = 1$ is the equation of a circle in the XY plane, whereas $X^2 - Y^2 = 1$ is the equation of a (rectangular) hyperbola. Indeed, cosine and sine are often referred to as the 'circular' functions ($X = \cos \theta$, $Y = \sin \theta$ being the parametric equation of a circle, whereas $X = \cosh \theta$, $Y = \sinh \theta$ is the parametric equation of a hyperbola).

The analogy between cosh and cos on the one hand, and sinh and sin on the other, also extends to the power series for these functions. To obtain the series expansion for $\cosh x$, you can use $\cosh x = (e^x + e^{-x})/2$ together with the series for e^x and e^{-x}. Another way to proceed is to observe that $(d^n/dx^n) \cosh x$ is just $\cosh x$ if n is even, and $\sinh x$ if n is odd; since $\cosh 0 = 1$ and $\sinh 0 = 0$ (another point of contact with the trigonometric functions!), you can then go ahead and write down a Taylor series about $x = 0$. However you do it, the result is the same. The series expansion for $\sinh x$ can be determined in the same way, and we find

$$\left.\begin{array}{l} \cosh x = 1 + \dfrac{x^2}{2!} + \dfrac{x^4}{4!} + \dfrac{x^6}{6!} + \cdots \\[2mm] \sinh x = x + \dfrac{x^3}{3!} + \dfrac{x^5}{5!} + \dfrac{x^7}{7!} + \cdots \end{array}\right\} \tag{7.9}$$

which are to be compared with

$$\left.\begin{array}{l} \cos x = 1 - \dfrac{x^2}{2!} + \dfrac{x^4}{4!} - \dfrac{x^6}{6!} \cdots \\[2mm] \sin x = x - \dfrac{x^3}{3!} + \dfrac{x^5}{5!} - \dfrac{x^7}{7!} \cdots \end{array}\right\} \tag{7.10}$$

The correspondence between equations (7.9) and (7.10) could hardly be more striking. Corresponding series are identical, apart from differences of sign. How are we to interpret these remarkable links between hyperbolic and trigonometric functions?

To understand more fully what is going on, we need to know a little of the algebra of **complex numbers**. Any pair a, b of real numbers defines a corresponding complex number $a + bi$, or $a + ib$, which is the same. Thus $2 + 3i$, $1 - \pi i$, $\sqrt{2} + i$ are all examples of complex numbers. Here i ($= 0 + 1.i$) is a very special complex number, assumed to satisfy the algebraic identity $i^2 = -1$. Any real number a may be regarded as a particular kind of complex number, $a = a + i.0$, for which $b = 0$.

Complex numbers can be

- added: $(2 + i) + (3 + 4i) = 5 + 5i$

- subtracted: $(2 + i) - (3 + 4i) = -1 - 3i$

- multiplied:
  $(1 + i) \times (2 + i) = 1 \times 2 + 1 \times i + i \times 2 + i \times i = 2 + i + 2i - 1 = 1 + 3i$

- divided: $\frac{1+i}{2+i} = \frac{(1+i)(2-i)}{(2+i)(2-i)} = \frac{2-i+2i-i^2}{4-2i+2i-i^2} = \frac{3+i}{5} = \frac{3}{5} + \frac{1}{5}i$

To carry out the operation of division, multiply both numerator and denominator by the **complex conjugate** of the denominator. The complex conjugate of a complex number z is usually denoted by $\bar{z}$, and is given for $z = a + ib$ by $\bar{z} = a - ib$; to divide z_1 by z_2 one has to evaluate $z_1/z_2 = z_1\bar{z}_2/z_2\bar{z}_2$, where the denominator will always be real.

The algebra of complex numbers works in a manner very similar to the algebra of real numbers, with the additional algebraic fact that $i^2 = -1$. Of course no real number x has the property that $x^2 = -1$. The complex number system may be regarded as an extension of the real number system, and the interested reader should consult one of the standard works on the subject to find out how this extension can be carried out. One of the basic results which can be proved is the so-called fundamental theorem of algebra, which states that if $p(z)$ is any polynomial in z of degree n, then the equation $p(z) = 0$ has exactly n complex roots (provided proper account is taken of double and multiple roots). We shall find this theorem of particular significance in the theory of differential equations.

Once the algebra of complex numbers has been established, a great deal of theory of functions can be extended to apply to complex as well as to real numbers, though to carry this out in detail would take us far beyond the scope of this book. In particular, the power series $\exp(z) = 1 + z + z^2/2! + z^3/3! + \dots$ makes sense if z is allowed to be complex, just as it does for real numbers. (Surprisingly, the theory of power series of complex numbers turns out to be a very powerful branch of analysis which throws considerable light on real power series as well.) The functions $\sin z$, $\cos z$, $\sinh z$ and $\cosh z$ may also be defined for complex z, through the simple expedient of writing down the same power series for z complex as is known already to hold for z real.

It is through allowing z to be complex in the exponential function that the link between exponential, hyperbolic and trigonometric functions is most clearly shown. Just take $z = ix$, with x real, and we have

$$e^{ix} = 1 + (ix) + \frac{(ix)^2}{2!} + \frac{(ix)^3}{3!} + \frac{(ix)^4}{4!} + \dots$$
$$= 1 + ix - \frac{x^2}{2!} - i\frac{x^3}{3!} + \frac{x^4}{4!} + \dots$$

which, on collecting together the terms without i and terms with i, yields

$$e^{ix} = \left(1 - \frac{x^2}{2!} + \frac{x^4}{4!} - \dots\right) + i\left(x - \frac{x^3}{3!} + \frac{x^5}{5!} - \dots\right)$$

On the right-hand side, we see the familiar series for cos x and sin x, and we have the result, known as De Moivre's Theorem,

$$e^{ix} = \cos x + i \sin x \qquad (7.11)$$

which holds for all $x \in \mathbb{R}$ (and can be extended to x complex as well!).

Equation (7.11) not only shows the link between exponential, cosine and sine, but can also be used to relate these functions to their hyperbolic counterparts. Substituting $-x$ for x in equation (7.11) gives the complex conjugate result $e^{-ix} = \cos x - i \sin x$, which on adding or subtracting the original equation leads to the identities

$$\left. \begin{array}{l} \cos x = \dfrac{e^{ix} + e^{-ix}}{2} = \cosh(ix) \\[2mm] \sin x = \dfrac{e^{ix} - e^{-ix}}{2i} = -i \sinh(ix) \end{array} \right\}$$

Hence the cosine and sine functions are expressible in terms of the cosh and sinh functions through simple substitutions; the corresponding formulae expressing cosh and sinh in terms of cos and sin are

$$\left. \begin{array}{l} \cosh x = \cos(ix) \\ \sinh x = -i \sin(ix) \end{array} \right\}$$

results which may easily be verified to agree with the series expansions obtained in equations (7.9) and (7.10).

I cannot resist concluding this section with a very famous algebraic consequence of De Moivre's Theorem. Set $x = \pi$ in equation (7.11) to obtain the identity $e^{i\pi} = -1$. A fundamental equation connecting three of the principle 'constants' of mathematics, e, i and π!

EXERCISES ON 7.4

1. Without further consulting this book, write down as much as you can about the exponential function, including a definition of this function, its domain and range, a sketch of its graph, and some of the principal identities satisfied by this function.

2. Use standard properties of the exponential function to prove the identities $\cosh(2x) = 2\cosh^2 x - 1$ and $\sinh(2x) = 2 \sinh x \cosh x$.

3. Assuming De Moivre's Theorem (equation (7.11)), as well as standard properties of the exponential function e^z, which may be supposed to hold for real as well as complex values of z, show that

$$e^{2ix} = (\cos^2 x - \sin^2 x) + 2i \sin x \cos x$$

Hence obtain the trigonometric identities

$$\cos 2x = \cos^2 x - \sin^2 x, \quad \sin 2x = 2 \sin x \cos x$$

By considering e^{3ix}, derive identities for $\cos 3x$ and $\sin 3x$.

4. Use standard properties of the exponential function, and the identities $\sin\theta = (e^{i\theta} - e^{-i\theta})/2i$, $\cos\theta = (e^{i\theta} + e^{-i\theta})/2$, to prove the trigonometric identity $\sin x \cos y + \cos x \sin y = \sin(x+y)$.

Summary

This chapter has been devoted to developing the properties of the exponential function, its inverse the natural logarithm, the hyperbolic functions, principally $\cosh x$ and $\sinh x$, and to using complex algebra to explore the links of these functions with the trigonometric sine and cosine. The best way to summarize the results of the chapter, and to remind the reader of some of the main ideas which will be carried further in later chapters of the book, will be to list the most relevant formulae satisfied by those functions which are new to this chapter, namely $\exp(x)$ $(= e^x)$, $\ln(x)$, $\cosh x$, $\sinh x$, and $\tanh x$.

- $\exp(x) = 1 + x + \frac{x^2}{2!} + \frac{x^3}{3!} + \frac{x^4}{4!} + ...$; $\exp(0) = 1$, $\exp(1) = e$

- $\cosh x = \frac{e^x + e^{-x}}{2} = 1 + \frac{x^2}{2!} + \frac{x^4}{4!} + \frac{x^6}{6!} + ...$; $\cosh(0) = 1$

- $\sinh x = \frac{e^x - e^{-x}}{2} = x + \frac{x^3}{3!} + \frac{x^5}{5!} + \frac{x^7}{7!} + ...$; $\sinh(0) = 0$

- $\tanh x = \sinh x / \cosh x$

- $\cosh^2 x - \sinh^2 x = 1$

- $\cosh 2x = 2\cosh^2 x - 1$; $\sinh 2x = 2 \sinh x \cosh x$

- $\ln(1 + x) = x - \frac{x^2}{2} + \frac{x^3}{3} - \frac{x^4}{4} + ...$ $(-1 < x \le 1)$; $\ln(1) = 0$, $\ln(e) = 1$

- $\frac{d}{dx}\exp(cx) = c\exp(cx)$; $\frac{d}{dx}\cosh x = \sinh x$; $\frac{d}{dx}\sinh x = \cosh x$; $\frac{d}{dx}\tanh x = \frac{1}{\cosh^2 x}$; $\frac{d}{dx}\ln(|x|) = 1/x$

- $\ln(ab) = \ln(a) + \ln(b)$; $\exp(A + B) = \exp A \times \exp B$

- $\ln(a/b) = \ln(a) - \ln(b)$; $\exp(A - B) = \exp A / \exp B$

- $-\ln(a) = \ln(1/a)$; $\exp(-A) = 1/\exp(A)$

- $\ln(\exp x) = x$; $\exp(\ln x) = x$

- $\log_a x = \ln(x)/\ln(a)$; $\frac{d}{dx}\log_a x = \frac{1}{x\ln(a)}$

- $x^c = \exp(c\ln(x))$

- $e^{ix} = \cos x + i \sin x$; $\sin x = \frac{e^{ix} - e^{-ix}}{2i}$; $\cos x = \frac{e^{ix} + e^{-ix}}{2}$; $\cos x = \cosh(ix)$; $\sin x = -i \sinh(ix)$; $\cosh x = \cos(ix)$; $\sinh x = -i \sin(ix)$

FURTHER EXERCISES

1. Differentiate each of the following functions: $\exp(-3x)$, $\exp(-x^2)$, $e^{2x}\cos 3x$, $x^2 e^{x^2}$, $\ln(3x)$, $\log_{10}(3x)$, $\ln(1 + x^2)$, $x^2(\ln(x))^2$, $x^x (= \exp(x\ln(x))$ for $x > 0$).
2. Assuming that $\ln(1 + x)$ has derivative 1 at $x = 0$, deduce that $\lim_{h\to 0}(\ln(1 + h))/h = 1$, and hence show that $\lim_{x\to\infty} x\ln(1 + (1/x)) = 1$. By considering the function $\ln(g(x))$, where $g(x) = (1 + (1/x))^x$, prove that $\lim_{x\to\infty}(1 + (1/x))^x = e$.

3. Simplify the expressions

$$\frac{d}{dx}\left\{\frac{(e^x - e^{-x})^2}{e^{2x}}\right\}, \exp(-3\ln(x)) \text{ and } \ln(x^2 + x^3)$$

4. Verify that

$$\frac{d^4}{dx^4}(e^x \sin x) = -4e^x \sin x$$

Use this result to determine

$$\frac{d^{4n}}{dx^{4n}}(e^x \sin x)$$

where n is any positive integer.

5. Find the maximum value of xe^{-2x}, for any positive value of x.

6. Verify the identity $\cosh(x + y) = \cosh x \cosh y + \sinh x \sinh y$.

7. Evaluate the following complex numbers: $(1 + i)^4$, $(1 + 2i)(1 + 3i)$, $1/(1 + i)$, $(1 - i)/(2i - 1)$, $\exp(i\pi/2)$.

8. A point in two-dimensional space has Cartesian coordinates (x, y) and polar coordinates (r, θ) where $x = r\cos\theta$, $y = r\sin\theta$. Use De Moivre's Theorem to show that the complex number $z = x + iy$ is given by $z = re^{i\theta}$. Find the polar coordinates for the point $x = 1$, $y = 1$, and hence evaluate $(1 + i)^{10}$.

9. If $\bar{z} = a - ib$ denotes the complex conjugate of the complex number $z = a + ib$, prove the identities $\overline{z_1 + z_2} = \bar{z}_1 + \bar{z}_2$, $\overline{z_1 z_2} = \bar{z}_1 \bar{z}_2$ and $\overline{(z_1/z_2)} = \bar{z}_1/\bar{z}_2$.

10. Starting from the identity $x^c = \exp(c\ln(x))$, valid for all $c \in \mathbb{R}$ and for $x > 0$, and assuming the derivative of the exponential and logarithmic functions, obtain the result

$$\frac{d}{dx}x^c = cx^{c-1}$$

11. Show that

$$\frac{d}{dx}\sinh^{-1}x = \frac{1}{\sqrt{1 + x^2}}$$

Show how an inverse cosh function may be defined, with domain $[1, \infty)$, and show that

$$\frac{d}{dx}\cosh^{-1}x = \frac{1}{\sqrt{x^2 - 1}} \quad \text{for } x > 1.$$

8 • The Antiderivative

I have emphasized in this book that differentiation should be seen as an operation which sends functions to functions. Seen in this light, it becomes very natural to think about the corresponding inverse operation. What I mean by this is that once we are thinking in terms of an operation, called differentiation, which sends functions f into their derivatives f', we can also consider an inverse operation, which I shall call antidifferentiation, which sends f' back again to f.

We know that $\cos x$ is the derivative of $\sin x$, so I shall call $\sin x$ an antiderivative of $\cos x$. More generally, an antiderivative of a function g is defined to be any function which differentiates to give g. A common notation which brings out quite clearly the relationship between a function and its antiderivative is that an antiderivative of g will often be denoted by G, an antiderivative of f by F, an antiderivative of h by H, and so on; so antiderivatives of functions f, g and h must satisfy, respectively, the equations $F' = f$, $G' = g$ and $H' = h$.

Notice that I have referred throughout to **an** antiderivative of a function, rather than **the** antiderivative. This is because a given function will have not just one antiderivative (assuming it has any at all!) but actually an infinite number of them. We know that $\sin x$ is an antiderivative of $\cos x$, because

$$\frac{\mathrm{d}}{\mathrm{d}x}\sin x = \cos x$$

but then so is $\sin x + c$ for any constant c, because we also have

$$\frac{\mathrm{d}}{\mathrm{d}x}(\sin x + c) = \cos x$$

In general, if F is an antiderivative of f, then so is $F + c$, for any constant c.

In this chapter, we shall examine the idea of antiderivative, together with the related notion of indefinite integral. We shall find that not every function has an antiderivative, and for those which do we shall see how to carry out the operation of antidifferentiation. Really what this amounts to is taking many of the previous ideas and methods of differentiation, and putting them into reverse!

8.1 Antiderivatives and the indefinite integral

In dealing with antiderivatives, it is sometimes necessary to be a little careful about domains. I shall always assume that the functions for which we seek antiderivatives are defined on **intervals**. With this proviso, the definition of antiderivative has already been given, but I shall state it again here for convenience.

• *Definition I*

A function F is said to be an antiderivative of a function f on an interval, if $F'(x) = f(x)$ for all x in the interval; that is, F is an antiderivative of f, on the interval, whenever f is the derivative of F, for x in the interval.

✳ *Example I*

Antiderivatives: $x^3/3$ is an antiderivative of x^2. Here the interval in question is the entire real line. Also $(x^3/3) + c$ is an antiderivative of x^2, for any constant c. To verify that $(x^3/3) + c$ is an antiderivative of x^2, just check that x^2 is the derivative of $(x^3/3) + c$ — this is what it means!

$-1/x$ is an antiderivative of $1/x^2$ on the interval $0 < x < \infty$, as too is $-(1/x) + c$; and $-1/x$ is also an antiderivative of $1/x^2$ on the interval $-\infty < x < 0$. We **cannot**, however, say that $-1/x$ is an antiderivative of $1/x^2$ on the entire real line $-\infty < x < \infty$, because we do **not** have $d(-1/x)/dx = 1/x^2$ for all $x \in \mathbb{R}$. The equation fails at $x = 0$. The function $1/x^2$ is not even defined at $x = 0$, and any interval on which we seek an antiderivative for this function must not contain $x = 0$.

Table 8.1 shows some antiderivatives of powers of x. In the case of positive integer powers of x, and the constant function 1, we have an antiderivative on the entire real line. For negative integer powers, we have to keep away from $x = 0$; we have, then, an antiderivative on the positive real line $0 < x < \infty$, or on the negative real line $-\infty < x < 0$.

A simple way to find an antiderivative for, say, $1/x^4$ is based on trial and error. It is fairly clear that a power of x is required, and since differentiation reduces the power by one, we try $1/x^3$. However,

$$\frac{d}{dx}\left(\frac{1}{x^3}\right) = -\frac{3}{x^4}$$

which is correct apart from the constant factor -3, so just divide $1/x^3$ by -3, and you soon find that

$$\frac{d}{dx}\left(-\frac{1}{3x^3}\right) = \frac{1}{x^4}$$

Table 8.1 Antiderivatives of powers of x.

Function	Antiderivative
x^3	$\frac{x^4}{4}$
x^2	$\frac{x^3}{3}$
x	$\frac{x^2}{2}$
1	x
$\frac{1}{x}$	$\ln(|x|)$
$\frac{1}{x^2}$	$-\frac{1}{x}$
$\frac{1}{x^3}$	$-\frac{1}{2x^2}$

as required. Hence $-1/3x^3$ is an antiderivative of $1/x^4$, say for $x > 0$, as also is $-(1/3x^3) + c$ for any constant c.

The table of antiderivatives may be extended to any integer power of x. An antiderivative of x^n, for any positive or negative integer n except $n = -1$, is $(x^{n+1})/(n+1)$. The special case $n = -1$, which fills a gap in the table, is quite different, and here an antiderivative is $\ln(x)$ for $x > 0$, and $\ln(-x)$ for $x < 0$.

A combination of trial and error with inspired guesswork will often lead to an antiderivative. Usually the key to finding an antiderivative is to know (or guess) the **form** of the antiderivative. Suppose, for example, you suspect that the function xe^{2x} has an antiderivative of the form $(ax + b)e^{2x}$. Well, then you want

$$\frac{\mathrm{d}}{\mathrm{d}x}\{(ax + b)e^{2x}\} = xe^{2x}$$

Here, the left-hand side is

$$(ax + b).2e^{2x} + ae^{2x} = 2axe^{2x} + (a + 2b).e^{2x}$$

Comparing this with what you want, namely $1.xe^{2x} + 0.e^{2x}$, the antiderivative will work provided you have $2a = 1$ and $a + 2b = 0$. This gives $a = \frac{1}{2}$ and $b = -\frac{1}{4}$, so an antiderivative of xe^{2x} is $(\frac{1}{2}x - \frac{1}{4})e^{2x}$. Note that, in seeking an antiderivative of xe^{2x} having the form $(ax + b)e^{2x}$, it is not supposed that **every** antiderivative of xe^{2x} is of this form; indeed $(ax + b)e^{2x} + c$ is an antiderivative for all $c \in \mathbb{R}$. You should just think of $(ax + b)e^{2x}$ in this case as a trial function.

Sometimes, if you have an off day in trying to decide the form that an antiderivative might take, your trial function may not work, and you have to try something else. An example of this follows.

✐ *Example 2*

Incorrect trial function for antiderivative: suppose you are looking for an antiderivative of the function $x \sin x$. A not unreasonable (though incorrect!) guess is to look for an antiderivative of the form $(ax + b) \sin x$. Then

$$\frac{\mathrm{d}}{\mathrm{d}x}\{(ax + b) \sin x\} = (ax + b) \cos x + a \sin x$$

This cannot possibly equal $x \sin x$, which has no cosine term. Try again! What about $(ax + b) \cos x$, which at least on differentiation will produce a term involving $x \sin x$? Well,

$$\frac{\mathrm{d}}{\mathrm{d}x}\{(ax + b) \cos x\} = -(ax + b) \sin x + a \cos x$$

This looks more likely! Since $x \sin x$, the function we want to arrive at, is $1.x \sin x + 0. \sin x + 0. \cos x$, a comparison of coefficients yields $-a = 1$, $b = 0$, and $a = 0$. What a disappointment! The constant a cannot be both 0 and -1! (and do not make the mistake that some students do in the heat of examinations, of setting $a = -1$ and $b = 0$ and conveniently, if hastily, forgetting about the second equation for a).

Let us have one last attempt. To be rather more sure to have all of the terms that we want, let us try for an antiderivative $(ax + b) \cos x + (cx + d) \sin x$. Probably not all of these terms will be necessary, but if so we are erring on the safe side, because those coefficients which we do not need will then turn out to be zero.

Now

$$\frac{d}{dx}\{(ax+b)\cos x + (cx+d)\sin x\} = (cx+d+a)\cos x - (ax+b-c)\sin x$$

For this to equal $(0x+0)\cos x + (1.x+0)\sin x$, we shall need to have $c=0$, $d+a=0$, $a=-1$, $b-c=0$. These four equations for the coefficients a, b, c, d actually **do** have a solution, namely $a=-1$, $d=1$, $b=c=0$. So we have, at last, found an antiderivative for the function $x\sin x$, namely $\sin x - x\cos x$.

To check, we have

$$\frac{d}{dx}(\sin x - x\cos x) = \cos x + x\sin x - \cos x$$

$$= x\sin x$$

as required. Success!

Of course, in this example, we could have saved a lot of time had we thought of the correct trial function from the start; and even more time, had we realized that the terms involving b and c in the latest trial function were unnecessary, since $b=c=0$, so we could have coped with a trial function $ax\cos x + d\sin x$. Certainly our success rate with trial functions (as with much else) will improve with experience, and to help out a bit further I shall give you some methods by which you can find antiderivatives in special cases. The next example looks at a few more cases for which it is relatively easy to find an antiderivative, and also presents a very common pitfall (or howler!) that you should try to avoid at all costs!

▣ *Example 3*

Further antiderivatives: it is required to find antiderivatives of each of the functions $\cosh 3x$, $\cos^2 x$, $1/(1+2x^2)$, $1/(x^2+2)$, $x/(x^2+1)$, $e^{2x}/(1+e^{2x})$, xe^{-x^2} and e^{-x^2}.

The first case is straightforward. We know that $\sinh$ differentiates to give $\cosh$, and

$$\frac{d}{dx}\sinh 3x = 3\cosh 3x$$

Adjusting the constant factor, an antiderivative is $\sinh 3x/3$.

For $\cos^2 x$ the situation is not so clear. However, the trigonometric identity $\cos^2 x = (1+\cos 2x)/2$ comes to our aid, and it is not difficult to find the antiderivative $(x/2)+(\sin 2x/4)$.

The function $f(x) = 1/(1+2x^2)$ looks difficult, but you may remember from Section 5.4 that the derivative of the inverse tangent function is given by

$$\frac{d}{dx}\tan^{-1}x = \frac{1}{1+x^2}$$

Here we have not x^2 but $2x^2$, which is $(x\sqrt{2})^2$, so a natural next step is to look at the derivative of $\tan^{-1}(x\sqrt{2})$. Using the chain rule, this comes out to be

$\sqrt{2}/(1+2x^2)$. Hence an antiderivative of the function $f(x) = 1/(1+2x^2)$ is $F(x) = (1/\sqrt{2})\tan^{-1}(x\sqrt{2})$.

For $f(x) = 1/(x^2 + 2)$, a similar argument may be applied, noting that $1/(x^2 + 2) = 1/2(1 + (x^2/2)) = 1/2(1 + (x/\sqrt{2})^2)$; this leads to $F(x) = (1/\sqrt{2})\tan^{-1}(x/\sqrt{2})$.

The case $x/(x^2 + 1)$ is rather different, and is a type of function which you should keep a look out for. The special feature of this function is that it has a numerator and a denominator, where the numerator x is (very much like) the derivative of the denominator $(1 + x^2)$. Any function of this form will have an antiderivative involving the logarithmic function. All that is necessary is to make use of the identity

$$\frac{\mathrm{d}}{\mathrm{d}x}\ln(h(x)) = h'(x)/h(x)$$

which is valid for $h(x) > 0$ and follows, for example, from the chain rule. In this example the numerator is just half the derivative of the denominator. Since

$$\frac{\mathrm{d}}{\mathrm{d}x}\ln(x^2 + 1) = \frac{2x}{x^2 + 1}$$

an antiderivative of $x/(x^2 + 1)$ will be $\frac{1}{2}\ln(x^2 + 1)$.

The next case, $f(x) = e^{2x}/(1 + e^{2x})$, is rather similar, since

$$\frac{\mathrm{d}}{\mathrm{d}x}(1 + e^{2x}) = 2e^{2x}$$

and an antiderivative is $\frac{1}{2}\ln(1 + e^{2x})$.

To find an antiderivative of xe^{-x^2}, an obvious step is to use a trial function containing e^{-x^2}. However, by the chain rule

$$\frac{\mathrm{d}}{\mathrm{d}x}e^{-x^2} = -2xe^{-x^2}$$

so $F(x) = -\frac{1}{2}e^{-x^2}$ will do.

For the final case, what about an antiderivative for the function e^{-x^2}? A **very common error** is to observe (correctly) that the derivative of e^{-x^2} is $-2xe^{-x^2}$, and to conclude (incorrectly) that all that is necessary is to adjust the antiderivative by dividing through by the factor $-2x$, to give $F(x) = -(1/2x)e^{-x^2}$ as an antiderivative. This is **totally incorrect!!** As you can see from carrying out the differentiation, using the product rule, which gives

$$\frac{\mathrm{d}}{\mathrm{d}x}\left(-\frac{1}{2x}e^{-x^2}\right) = -\frac{1}{2x}\frac{\mathrm{d}}{\mathrm{d}x}e^{-x^2} + e^{-x^2}\frac{\mathrm{d}}{\mathrm{d}x}\left(-\frac{1}{2x}\right)$$

$$= -\frac{1}{2x}(-2xe^{-x^2}) + e^{-x^2}\left(\frac{1}{2x^2}\right) = e^{-x^2}\left(1 + \frac{1}{2x^2}\right)$$

not at all what was required! The problem is, of course, that to divide a function by $-2x$ **before** differentiation will in no way compensate for an unwanted factor $-2x$ **after** differentiation. The fact that

$$\frac{\mathrm{d}}{\mathrm{d}x}(e^{-x^2}) = -2xe^{-x^2}$$

does **not** allow us to conclude

$$\frac{d}{dx}\left(\frac{e^{-x^2}}{-2x}\right) = -\frac{1}{2x}(-2xe^{-x^2})$$

and the product rule tells us so, in no uncertain terms. (Such an argument does, however, work if only multiplication by **constants** is involved. So, for example,

$$\frac{d}{dx}(e^{-x^2}) = -2xe^{-x^2}$$

does allow us to conclude that

$$\frac{d}{dx}\left(-\frac{1}{2}e^{-x^2}\right) = xe^{-x^2}$$

which is the argument which we used successfully with the previous antiderivative.)

I would like to emphasize here that the only really safe way to confirm that a function F is an antiderivative of a function f is to carry out the differentiation, and that this must be done using the correct rules of differentiation, as I have set them out in Chapter 5. If you are uncertain, simply verify that $d(F(x))/dx = f(x)$.

If $e^{-x^2}/-2x$ is not a correct antiderivative of the function e^{-x^2}, how in fact can a true antiderivative be found? In fact, this case is not so straightforward as it might appear. The function $f(x) = e^{-x^2}$ is one of a number of functions apparently simple in form, and involving only standard functions about which we have a great deal of knowledge, for which there is **no** simple expression for an antiderivative in terms of our basic armoury of functions. It is not that $f(x) = e^{-x^2}$ has **no** antiderivative. In fact, it may be shown that every continuous function on an interval has an antiderivative on that interval, and certainly e^{-x^2} is an example of a continuous function. There **is** an antiderivative in this case, and, for reasons which I shall not go into here, involving the so-called normal distribution of probability theory, it has been given the name of error function; but the error function cannot be expressed in terms of standard functions such as polynomials, powers, the exponential, logarithm, trigonometric functions, and so on. If we are just restricted to these functions, then we have no hope of finding an antiderivative, by trial and error or by any other means.

To convince yourself that an antiderivative does exist in this case, you will need to resort to power series expansions. Substituting $-x^2$ into the power series for the exponential function, we have $f(x) = e^{-x^2} = 1 - x^2 + (x^4/2!) - (x^6/3!) + (x^8/4!) - ...$, so that a suitable antiderivative is given by $F(x) = x - (x^3/3) + (x^5/5.2!) - (x^7/7.3!) + (x^9/9.4!) - ....$

As may be verified by the ratio test, this series for $F(x)$ converges for all $x \in \mathbb{R}$, and hence has infinite radius of convergence, so the series may properly be differentiated term by term to obtain the result $d(F(x))/dx = f(x)$, as required. Of course, from what I have said, it follows that the summation of the series for $F(x)$ cannot be carried out, in the sense of expressing the sum in terms of standard functions that we know all about.

In a sense, what is true in the case of an antiderivative for the function e^{-x^2} holds also to some degree for some of the functions with which we are most familiar. There is no way in which the power series for the exponential function, or the sine and cosine series for that matter, can be summed 'exactly'. What number is the sine of one radian? We can calculate this number to any desired accuracy, but a precise evaluation is beyond us; nor is this number expressible in a simple way in terms of other, more amenable functions. But mathematicians have given names to the exponential function, the sine and cosine, the error function, and numerous other functions with which we can work quite well by being aware of their general properties, and in the absence of anything that would allow us to determine their values except to some degree of approximation. We should be content with this situation, which prevails throughout almost the whole of mathematics in its many applications.

I hope that by now you will have a good idea of how to obtain antiderivatives in some of the simpler cases. We shall extend this knowledge in some of the later sections of this chapter, by looking at some quite powerful methods for finding antiderivatives.

I have already referred to the fact that if $F(x)$ is an antiderivative of $f(x)$, then so too will be $F(x) + c$, for any constant c. In fact, **any** two antiderivatives of a given function $f(x)$, on an interval, can only differ by a constant. This is because if F_1 and F_2 are two antiderivatives on the interval, then $\mathrm{d}F_1/\mathrm{d}x = f(x)$ and $\mathrm{d}F_2/\mathrm{d}x = f(x)$ together imply that

$$\frac{\mathrm{d}}{\mathrm{d}x}(F_1(x) - F_2(x)) = f(x) - f(x) = 0$$

for all x belonging to the interval, and this can only be so if $F_1(x) - F_2(x) = c$ is a constant function. Hence if you have found just one antiderivative $F(x)$ of $f(x)$ then you have a complete knowledge of **all** antiderivatives on the interval. The general antiderivative of f will then be $F(x) + c$, for any $c \in \mathbb{R}$.

The general expression for an antiderivative of a given function $f(x)$ on an interval is called **the indefinite integral** of $f(x)$, on the interval, and will be denoted by $\int f(x)\mathrm{d}x$. If, instead of x, functions are expressed in terms of an input value t, say, then the indefinite integral will be written $\int f(t)\mathrm{d}t$. The indefinite integral of a function is then not a function but a set of functions, any two functions of the set differing by a constant. This reflects the fact that the operation of differentiation is not **injective**. Differentiation can send two different functions, provided they differ by a constant, into the same derivative function. As we know from our experience with inverses, if an operation is non-injective then there will be no unique inverse operation — for this reason the indefinite integral defines a collection of functions, given by $F(x) + c$ for all possible $c \in \mathbb{R}$. This, in fact, is why we speak of an **indefinite** integral. The definite integral, which is closely related but which is defined in terms of area, will be considered in Chapter 9.

All of the examples of antiderivatives that we have discussed so far may be expressed in the notation of indefinite integrals. We have, for example, $\int x^n \mathrm{d}x = (x^{n+1})/(n+1) + c$ $(n \neq -1)$, and $\int (1/x)\mathrm{d}x = \ln(|x|) + c$, $\int [e^{2x}/(1+e^{2x})]\mathrm{d}x = \frac{1}{2}\ln(1+e^{2x}) + c$, and so on. The main lesson of this notation is that, having found an antiderivative $F(x)$, the most general expression for a function which differentiates to give $f(x)$ is not $F(x)$ but $\int f(x)\mathrm{d}x = F(x) + c$.

EXERCISES ON 8.1

1. Find an antiderivative of the function $e^x \cos 2x$, having the form $e^x(A \cos 2x + B \sin 2x)$, where A and B are constants to be determined. What form do you expect an antiderivative of $x^2 e^x$ to take? Hence evaluate the indefinite integral $\int x^2 e^x dx$.
2. If F is an antiderivative of f, show that the function $F(\sin x)$ is an antiderivative of $\cos x.f(\sin x)$. Use this result to obtain the antiderivative $\sin^4 x/4$ of $\sin^3 x \cos x$. Obtain also antiderivatives of the functions $\cot x$ $(0 < x < \pi)$, and $\cos^3 x$, given that $\cot x = \cos x/\sin x$ and $\cos^3 x = \cos x(1 - \sin^2 x)$.
3. Verify that

$$\frac{d}{dx}(x \ln(x)) = 1 + \ln(x), \text{ for } x > 0$$

Hence evaluate $\int \ln(x) dx$ for $x > 0$. Determine also the indefinite integral $\int x \ln(x) dx$ for $x > 0$.
4. Evaluate $\int \frac{x^3}{1+x^4} dx$ and $\int \frac{1+x}{1+x^2} dx$.
5. Using the result of exercise 11 at the end of Chapter 7, that

$$\frac{d}{dx} \sinh^{-1} x = \frac{1}{\sqrt{1+x^2}}$$

find antiderivatives of the functions

$$\frac{1}{\sqrt{4+x^2}} \text{ and } \frac{1}{\sqrt{x^2 + 2x + 2}}$$

6. A function $g(t)$ is defined by

$$g(t) = 1 - t, \quad 0 \le t < 1$$
$$= 1, \quad 1 \le t \le 2$$

Verify that g has **no** antiderivative on the interval $0 \le t \le 2$.
[Hint: Any antiderivative would have to be of the form

$$G(t) = t - \frac{t^2}{2} + c_1, \quad 0 \le t < 1$$
$$= t + c_2, \quad 1 \le t \le 2$$

where $c_2 = c_1 - \frac{1}{2}$ since G has to be continuous at $t = 1$. Show that G is not differentiable at $t = 1$.]

8.2 Uses of the antiderivative

The antiderivative is involved whenever we have to convert information about rate of change of a quantity into information about the quantity itself.

Let us suppose, for example, that we have a lake or reservoir which contains a volume u, of water, and that the volume of water u in the reservoir changes with time. Suppose, further, that we know the **rate of change** q of the volume of water in the reservoir, as a function of time. We might have this information, for example, in the form of data about the precipitation of rain into the lake, or the outflow/inflow of water via streams leading into the lake. Rainfall data in the form of centimetres per hour can be used, for example, to determine the contribution to rate of change of volume, once the total surface area of the lake is known. Certainly the precipitation rate will vary with time. So $u = u(t)$, $q = q(t)$, and the functions u and q are related by the equation $du(t)/dt = q(t)$. Hence $u(t)$ must be an antiderivative of $q(t)$, and we can write $u(t) = \int q(t)dt + c$, where c is a constant. The constant c is undetermined until we are given one further piece of information, the volume of water in the lake at some initial time, say $t = 0$. We cannot know in absolute terms how much water is in the lake at some given time t if all the information we have concerns how the total volume **changes** with time. Indeed, it may introduce a whole range of new practical problems to arrive at a realistic figure for the volume of a lake. There are many deep lakes which have not been fully charted below the surface, and for which no reliable figures on the depth are available. In that case our conclusion that $u(t) = \int q(t)dt + c$ still stands, in the absence of a precise value for the constant c.

There is another problem which we may consider here. Suppose that the principal cause of change in volume of water in our reservoir is outflow of water, and that the rate of outflow is known as a function of water depth and therefore indirectly as a function of volume. It is quite a reasonable hypothesis to suppose that as water volume and depth increase, so does the rate of outflow across barriers or into streams, and we may wish to model the rate of change of volume q by setting $q = q(u)$, a given function of volume u.

In this case, we have $du(t)/dt = q(u)$. Here of course the right-hand side function $q(u)$ does depend on t through the equation $u = u(t)$, and it is possible to write $u(t) = \int q(u(t))dt$. However, we cannot actually evaluate this antiderivative, since to do so we would need first of all to specify $q(u(t))$ as a function of t, and this we cannot do until we know the function $u(t)$; and this, of course, is what we are trying to determine! (I hope you will **not**, here, fall into the trap of writing $\int q(u)dt = (q^2/2) + c$. This is completely incorrect, because by the chain rule $(d/dt)((q^2/2) + c)$ is **not** q but qdq/dt.)

How then, can we determine the function $u(t)$, if it is known that $du/dt = q(u)$ for some given function u? I am straying here a little into the area of differential equations, which will concern us particularly in Chapter 10. However, since the answer to this problem is so closely linked to our current concern of antiderivatives, it may be helpful to give at least some preliminary attention to this question now.

The equation $du/dt = q(u)$ may be written in a more amenable form by taking the reciprocal of both sides, and writing $dt/du = 1/q(u)$. Here the right-hand side is a known function of u, so that $t = \int (1/q(u))du + c$. Although this approach seems quite promising, it does need a little care. For example, we know that u is a function of t, but it may not be so clear that the dependence of u on t can be inverted to express t as a function of u; we know from our discussion of inverse functions that this can be done if $u = u(t)$ is injective, but this is an assumption that we might need to verify.

In fact, the procedure which we have adopted here can usually be justified. Assuming $q(u) \neq 0$, we can define a function $f(u) = 1/q(u)$. As long as we can find an antiderivative $F(u) = \int (1/q(u))\mathrm{d}u$, on some interval of values of u, then with $u = u(t)$, the chain rule tells us that

$$\frac{\mathrm{d}}{\mathrm{d}t}F(u) = \frac{\mathrm{d}F(u)}{\mathrm{d}u}\cdot\frac{\mathrm{d}u}{\mathrm{d}t} = \frac{1}{q(u)}\frac{\mathrm{d}u}{\mathrm{d}t} = 1$$

since $\mathrm{d}u/\mathrm{d}t = q(u)$.

Since $\mathrm{d}F(u)/\mathrm{d}t = 1$, we have $F(u) = \int(1/q(u))\mathrm{d}u = t + \text{const}$, which is the conclusion which we came to earlier. Although this result is not quite what we wanted, in that it expresses t as a function of u rather than u as a function of t, this may be all that is really needed, and it may in any case be possible to invert and determine the function $u = u(t)$ explicitly, for each constant c. In any case, the main point which I wish to make is that here again the antiderivative is the key to any successful approach to the problem.

In dynamics too, quite apart from their use in the solution of differential equations, antiderivatives are central to the foundations of the subject. For a particle moving along the x-axis, and subject to a force $f = f(x)$ which depends only on the position of the particle, through its coordinate x, the 'equation of motion' is the differential equation force equals mass times acceleration, or

$$m\frac{\mathrm{d}^2 x}{\mathrm{d}t^2} = f(x)$$

One then defines a potential energy function $V = V(x)$, up to an additive constant, to be an antiderivative of $-f$. Thus $\mathrm{d}V(x)/\mathrm{d}x = -f(x)$, or $f(x) = -\mathrm{d}V(x)/\mathrm{d}x$. In the case of simple harmonic motion, for example, in which $f(x) = -kx$, one has $V(x) = (kx^2/2) + \text{const}$.

The point of this definition lies in a remarkable identity satisfied by the potential energy $V(x)$, which follows from the equation of motion, together with the chain rule. With $x = x(t)$, $\mathrm{d}x/\mathrm{d}t = v(t)$, we have

$$\frac{\mathrm{d}}{\mathrm{d}t}V(x) = \frac{\mathrm{d}V(x)}{\mathrm{d}x}\frac{\mathrm{d}x}{\mathrm{d}t} = -f(x)\frac{\mathrm{d}x}{\mathrm{d}t} = -m\frac{\mathrm{d}^2 x}{\mathrm{d}t^2}\frac{\mathrm{d}x}{\mathrm{d}t} = -m\frac{\mathrm{d}v}{\mathrm{d}t}.v$$

where v is the velocity of the particle; but $-m(\mathrm{d}v/\mathrm{d}t).v$ is just $\mathrm{d}(-\frac{1}{2}mv^2)/\mathrm{d}t$.

Hence, we have shown that

$$\frac{\mathrm{d}}{\mathrm{d}t}V(x) = -\frac{\mathrm{d}}{\mathrm{d}t}(\tfrac{1}{2}mv^2)$$

or equivalently

$$\frac{\mathrm{d}}{\mathrm{d}t}(\tfrac{1}{2}mv^2 + V(x)) = 0$$

This implies that $\frac{1}{2}mv^2 + V(x)$ is constant, that is independent of t. We have already described $V(x)$, an antiderivative of minus the force, as the potential energy of the particle; and $\frac{1}{2}mv^2$, for a particle of mass m and velocity v, is the kinetic energy. The equation $\frac{1}{2}mv^2 + V(x) = \text{const}$ then states the principle of conservation of energy for a particle moving in one dimension in a field of force — the total energy of the particle, that is kinetic energy plus potential energy, remains constant throughout the motion.

Conservation of energy holds, too, in greater generality for a particle moving in a field of force in three or more dimensions.

EXERCISES ON 8.2

1. For a particle moving in one dimension and acted upon by a force $f(x)$, and regarding the velocity $v = dx/dt$ as a function of position x, use the chain rule to show that $mv\,dv/dx = f(x)$. Defining a potential energy function V by $f(x) = -dV(x)/dx$, verify that

$$\frac{d}{dx}\left(\tfrac{1}{2}mv^2 + V(x)\right) = 0$$

and hence that total energy of the particle is independent of its position.

2. A point P is at the origin at time $t = 0$, and moves along the x-axis with velocity dx/dt given by $dx/dt = x$. Show that $x(t) = e^t$
   (a) by considering $\frac{d}{dt}(e^{-t}.x)$ and using the product rule;
   (b) by considering $\frac{d}{dt}(\ln(x))$ and using the chain rule.

3. A cyclist sets off in a race at time $t = 0$, and for the first few seconds of the race has speed given as a function of distance travelled s by $ds/dt = k\sqrt{s}$, where k is a constant. By considering $d(\sqrt{s})/dt$, determine s as a function of time.

4. The volume u of water in a lake is governed by the equation $du/dt = -ku^{1/3}$, where k is a positive constant. By considering $d(u^{2/3})/dt$, determine how long the lake will take to run dry, given that initially there is a volume u_0 of water in the lake.

8.3 Methods of integration

On a given interval, any two antiderivatives of a given function will differ by a constant. It is then only necessary to find one antiderivative $F(x)$, in order to be able to write down the general antiderivative, or indefinite integral, $\int f(x)dx = F(x) + c$.

In the case of a function such as $f(x) = 1/(1 - x^2)$, which is not defined at the points $x = \pm 1$, one has to find antiderivatives on each of the three intervals $-\infty < x < -1$, $-1 < x < 1$, and $1 < x < \infty$, unless x has been specified to lie in one or other of these intervals. Sometimes, as in the case $\int(1/x)\,dx = \ln(|x|) + c$, a single expression can encompass the antiderivative on each of the prescribed intervals; here we have really a shorthand for writing $\int(1/x)\,dx = \ln(x) + c_1$ $(x > 0)$, and $\int(1/x)\,dx = \ln(-x) + c_2$ $(x < 0)$, where the respective constants c_1 and c_2 relate to different intervals, and have nothing to do with each other.

The process of finding all of the antiderivatives of a given function is the process, or operation, of integration. Integration, the determination of indefinite integrals, is the inverse operation to differentiation. Any general property of the derivative will also tell us something about the integral. It is therefore not particularly surprising that the three methods of integration which I shall present here, namely (i) integration by parts, (ii) integration by substitution or change of variable, and (iii) integration by the method of partial fractions, are in turn closely connected with results or ideas that I have already talked about with respect to differentiation;

these are (i) the product rule, (ii) the chain rule, and (iii) the linearity of differentiation.

You may have noticed, perhaps in our treatment of the function e^{-x^2} in Section 8.1, a limitation in what can be achieved in the way of integration, which does not apply in the case of differentiation. You may, by now, feel confident enough to attempt the differentiation of just about **any** function which can be expressed in a simple way in terms of powers, exponentials, trigonometric functions and all the other functions which you use daily in mathematics theory and applications. You may not feel so confident with regard to integration. There are quite simple functions, of which e^{-x^2} is one, but $\cos(x^4)$, $\sin x / x$ and $\sqrt{\ln(x)}$ are others which can be differentiated quite easily, but which have no integral which can be expressed in closed form in terms of the basic functions of calculus. On any interval on which these functions are defined and continuous, the indefinite integral **does** exist; it is just that the integral can only be expressed as a limit that cannot be evaluated in closed form, or as a series for which the sum is unknown, and so on. Nevertheless, I hope that you will not be deterred from approaching integration in a positive spirit. It is fundamental to the subject. It may be more difficult than differentiation, but the methods to follow should go some way towards making the task an easier one.

Integration by parts

The method of integration by parts is based on the product rule for differentiation, namely

$$\frac{d}{dx}(fg) = f\frac{dg}{dx} + g\frac{df}{dx}$$

and uses the simple expedient of transferring the second term on the right-hand side to the left-hand side. Then

$$\frac{d}{dx}(fg) - g\frac{df}{dx} = f\frac{dg}{dx} \tag{8.1}$$

Suppose that we wish to find an antiderivative of the product $f\,dg/dx$, and let us denote by $\int g(df/dx)\,dx$ an antiderivative of the product $g\,df/dx$. These two antiderivatives are related, because we have

$$\frac{d}{dx}\int g\frac{df}{dx}dx = g\frac{df}{dx}$$

implying that

$$\frac{d}{dx}\left\{fg - \int g\frac{df}{dx}dx\right\} = \frac{d}{dx}(fg) - \frac{d}{dx}\int g\frac{df}{dx}dx = \frac{d}{dx}(fg) - g\frac{df}{dx} = f\frac{dg}{dx}$$

from equation (8.1). This means that an antiderivative of $f\,dg/dx$ is $fg - \int g\,(df/dx)\,dx$, and we can write

$$\int f\frac{dg}{dx}dx = fg - \int g\frac{df}{dx}dx \tag{8.2}$$

Equation (8.2) is the formula for integration by parts. It does not solve all our problems for us, because it simply expresses the indefinite integral of one product, $f\,\mathrm{d}g/\mathrm{d}x$, in terms of the indefinite integral of another product, $g\,\mathrm{d}f/\mathrm{d}x$; and maybe the second product, $g\,\mathrm{d}f/\mathrm{d}x$, is even more difficult to integrate than the first. But, used in the right way, the integration by parts formula is an extremely useful tool in the evaluation of integrals, particularly where the integral of a product of two functions is required. We can **always** differentiate the product of two functions (provided only that we know how to differentiate each of the functions themselves!). We **cannot** always integrate the product of two functions, but often we can! Here are some examples showing how it can be done.

● *Example 4*

Integration by parts: we wish to use integration by parts to evaluate the indefinite integral $\int x^2 e^x \mathrm{d}x$.

The integration by parts formula is for integrating a product of the form $f\,\mathrm{d}g/\mathrm{d}x$, so in this example we have to express the 'integrand' $x^2 e^x$ as a product of two functions, of which one has to be f and the other $\mathrm{d}g/\mathrm{d}x$. The obvious way to do this seems to be with $f = x^2$ and $g = e^x$, but other choices are possible. We could, for example, take $f = x$ and $\mathrm{d}g/\mathrm{d}x = xe^x$, or even $f = 1$ and $\mathrm{d}g/\mathrm{d}x = x^2 e^x$, or $f = e^x$ and $\mathrm{d}g/\mathrm{d}x = x^2$. With experience, you will usually see quite quickly how to carry out the factorization. There are good reasons, which I shall explain shortly, why it is better to choose $f = x^2$ and $\mathrm{d}g/\mathrm{d}x = e^x$ than any of the other possibilities, so for the moment let us make this choice and see how to proceed.

Take $f(x) = x^2$ and $\mathrm{d}g(x)/\mathrm{d}x = e^x$. Rather than f and $\mathrm{d}g/\mathrm{d}x$ we need to know f and g, which is quite easy in this case because we can set $f(x) = x^2$ and $g(x) = e^x$. There is no need in this situation to write $g(x) = e^x + c$, because to carry out integration by parts you have to make a **choice** of f and g such that $f\,\mathrm{d}g/\mathrm{d}x$ is your given integrand, and to add a constant to g would only complicate matters. With $f = x^2$ and $g = e^x$, the integration by parts formula

$$\int f \frac{\mathrm{d}g}{\mathrm{d}x}\mathrm{d}x = fg - \int g \frac{\mathrm{d}f}{\mathrm{d}x}\mathrm{d}x$$

becomes

$$\int x^2 e^x \mathrm{d}x = x^2 e^x - \int e^x \frac{\mathrm{d}}{\mathrm{d}x}(x^2)\mathrm{d}x = x^2 e^x - \int 2xe^x \mathrm{d}x$$

Here we have made progress! The function $2xe^x$ is easier to integrate than our original function $x^2 e^x$. You may know the integral of this function already, in which case we are finished. If not, one more integration by parts should do the job. To integrate $2xe^x$, set $f = 2x$ and $\mathrm{d}g/\mathrm{d}x = e^x$. We can take $f = 2x$ and $g = e^x$, so that the integral becomes

$$\int 2xe^x \mathrm{d}x = 2xe^x - \int e^x \frac{\mathrm{d}}{\mathrm{d}x}(2x)\mathrm{d}x = 2xe^x - \int 2e^x \mathrm{d}x$$

Since the indefinite integral of $2e^x$ is $2e^x + \text{const}$, our original integral now becomes $\int x^2 e^x \mathrm{d}x = x^2 e^x - 2xe^x + 2e^x + c$. Hence integration by parts has given us an antiderivative of $x^2 e^x$, namely $(x^2 - 2x + 2)e^x$.

Let us now return to the question of how to decide upon the best factorization $x^2e^x = f\,dg/dx$. The first thing to realize is that whatever you do, you will have to carry out an integration to determine g, once f and dg/dx have been decided upon. Certainly $f = 1$ and $dg/dx = x^2e^x$ would be a particularly unfortunate choice, since to determine g in that case would require you to carry out the very integral that you wish to evaluate in the first place; and $f = x$, $dg/dx = xe^x$ would not be much better. The choice $f = e^x$, $dg/dx = x^2$ looks all right, with $f = e^x$ and $g = x^3/3$, but in that case the integration by parts formula throws up an integral, namely $\int (x^3/3)e^x dx$ which is actually **more** difficult to deal with than the original! So all in all, $f = x^2$, $dg/dx = e^x$ seems about the best choice available.

Here are two pieces of advice which may help you with future examples of integration by parts. The first is, why not make a choice (or guess) of g first of all, then evaluate dg/dx and see what f has to be in order to give you the required product $f\,dg/dx$. That way, no integration is required to obtain g from dg/dx, because you have gone the other way round and obtained dg/dx from g. The second piece of advice is to remember that integration by parts converts an integral of one function f multiplied by the derivative of another g into an integral of one function g multiplied by the derivative of another f. In the course of integration, the derivative shifts from g on to f, and the key question is, to evaluate the integral, would you rather see the derivative on f or on g? If you would rather see the derivative on f, then you are on the right lines in using integration by parts to evaluate $\int f(dg/dx)\,dx$. Otherwise not, because then you will end up with an integral more difficult than the original.

❧ *Example 5*

Consider now the similar but slightly more complicated example $\int x^3e^{2x}dx$. Following my advice of deciding on g first of all, try $g(x) = e^{2x}$, so that $dg/dx = 2e^{2x}$, and $f\,dg/dx = x^3e^{2x}$ gives $f(x) = x^3/2$. Then integration by parts gives

$$\int \frac{x^3}{2}\cdot(2e^{2x})dx = \int \frac{x^3}{2}\frac{d}{dx}(e^{2x})dx$$

$$= \frac{x^3}{2}e^{2x} - \int e^{2x}\frac{d}{dx}\left(\frac{x^3}{2}\right)dx$$

$$= \frac{x^3}{2}e^{2x} - \int \frac{3}{2}x^2e^{2x}dx$$

This is $(x^3/2)e^{2x} - \frac{3}{2}\int x^2e^{2x}dx$, and carrying on in the same spirit leads to

$$\int x^3e^{2x}dx = \frac{x^3}{2}e^{2x} - \frac{3}{2}\left\{\frac{x^2}{2}e^{2x} - \int e^{2x}\cdot\frac{d}{dx}\left(\frac{x^2}{2}\right)dx\right\}$$

$$= \frac{x^3}{2}e^{2x} - \frac{3}{4}x^2e^{2x} + \frac{3}{2}\int xe^{2x}dx$$

which with one further integration by parts yields

$$\int x^3 e^{2x} dx = \frac{x^3}{2} e^{2x} - \frac{3}{4} x^2 e^{2x} + \frac{3}{2} \left\{ \frac{x}{2} e^{2x} - \int \frac{1}{2} e^{2x} dx \right\}$$

$$= \left(\frac{x^3}{2} - \frac{3}{4} x^2 + \frac{3}{4} x - \frac{3}{8} \right) e^{2x} + c$$

It does no harm to verify the final result by differentiation, and this we do using the product formula, to give

$$\frac{d}{dx} \left\{ \left(\frac{x^3}{2} - \frac{3}{4} x^2 + \frac{3}{4} x - \frac{3}{8} \right) e^{2x} \right\} = \left(\frac{x^3}{2} - \frac{3}{4} x^2 + \frac{3}{4} x - \frac{3}{8} \right) .2e^{2x}$$

$$+ e^{2x} \left(\frac{3x^2}{2} - \frac{3}{2} x + \frac{3}{4} \right)$$

which happily reduces to $x^3 e^{2x}$ as required.

✸ *Example 6*

The two examples $\int x^3 e^{-x^2} dx$ and $\int x^2/(1+x^2)^2 \, dx$ require a slightly greater degree of ingenuity for their integration. In the first case, the factorization of the integrand $f = x^3$, $dg/dx = e^{-x^2}$ leads nowhere, because as we saw in Section 8.1 the function e^{-x^2} has no simple antiderivative. However, we can make things much easier for ourselves, as I said earlier, by deciding on g rather than dg/dx first of all. And what could be a more hopeful choice than to take $g(x) = e^{-x^2}$? The other factor in the integrand will then involve a power of x, and certainly it would be a move in the right direction to transfer the derivative from the exponential to the power. So, undeterred, we try $g(x) = e^{-x^2}$, in which case $dg/dx = -2xe^{-x^2}$, so that $f dg/dx = x^3 e^{-x^2}$ gives $f(x) = -x^2/2$. Integration by parts then leads to

$$\int -\frac{x^2}{2} \frac{d}{dx} (e^{-x^2}) dx = -\frac{x^2}{2} e^{-x^2} - \int e^{-x^2} \frac{d}{dx} \left(-\frac{x^2}{2} \right) dx$$

$$= -\frac{x^2}{2} e^{-x^2} + \int x e^{-x^2} dx$$

This is certainly a great simplification, in that the power in the integrand has been reduced from x^3 to x. In fact, things are even better than that, because we have just seen that

$$\frac{d}{dx} (e^{-x^2}) = -2xe^{-x^2}$$

so an antiderivative of xe^{-x^2} is $-\frac{1}{2} e^{-x^2}$, and our task is complete,

$$\int x^3 e^{-x^2} dx = -\frac{x^2}{2} e^{-x^2} - \frac{1}{2} e^{-x^2} + c$$

$$= -\frac{1}{2} (1 + x^2) e^{-x^2} + c$$

Again, it is a useful check to verify this result by differentiation.

The second integration, $\int x^2/(1+x^2)^2 \, dx$, is somewhat similar. The factor $1/(1+x^2)^2$ in the integrand, with the chain rule and the fact that

$d(-1/t)/dt = 1/t^2$, may be a clue that we should look at something like $g(x) = 1/(1+x^2)$. In that case, $dg/dx = -2x/(1+x^2)^2$ and $f\,dg/dx = x^2/(1+x^2)^2$ requires $f(x) = -\frac{1}{2}x$. Integration by parts then gives

$$\int -\frac{1}{2}x\frac{d}{dx}\left(\frac{1}{1+x^2}\right)dx = -\frac{1}{2}\frac{x}{1+x^2} - \int \frac{1}{1+x^2}\frac{d}{dx}\left(-\frac{1}{2}x\right)dx$$

$$= -\frac{1}{2}\frac{x}{1+x^2} + \frac{1}{2}\int \frac{1}{1+x^2}\,dx$$

From our discussion in Section 5.4 of the differentiation of inverse trigonometric functions, you may recall that

$$\frac{d}{dx}\tan^{-1}x = \frac{1}{1+x^2}$$

(However, if you do not, then our next method of integration will probably help you.) The integration is now complete, and we have

$$\int \frac{x^2}{(1+x^2)^2}\,dx = -\frac{1}{2}\frac{x}{1+x^2} + \frac{1}{2}\tan^{-1}x + c$$

❋ *Example 7*

We know how to differentiate $\ln(x)$. How do we integrate this function? At first sight on considering the integral (for $x > 0$) $\int \ln(x)dx$, there does not seem much scope for factorization into a product of f with dg/dx. However, we can get round this problem by manufacturing a product where none appeared to exist. This we may do by setting $f = \ln(x)$, $dg/dx = 1$. (The choice $f = 1$, $dg/dx = \ln(x)$ will quickly be found not to lead anywhere.) Then $f = \ln(x)$, $g = x$, and integration by parts gives us

$$\int \ln(x)\frac{d}{dx}(x)dx = \ln(x).x - \int x\frac{d}{dx}\ln(x)dx$$

$$= x\ln(x) - \int x.\frac{1}{x}dx$$

$$= x\ln(x) - \int 1dx$$

$$= x\ln(x) - x + c$$

This example is rather easier than it looked at first sight, and a similar approach can often be successful in integrating functions involving the logarithm.

Integration by substitution

Integration by substitution, or change of variable, is closely related to the use of the chain rule for differentiation. It will be helpful to consider first of all a simple example.

To apply the chain rule to the derivative

$$\frac{d}{dx}\frac{1}{1+x^2}$$

we break the problem up by writing $y = 1/t$, $t = 1 + x^2$. Then $dy/dt = -1/t^2$, $dt/dx = 2x$, and the chain rule gives us $dy/dx = (dy/dt).(dt/dx) = -(1/t^2).2x$, which on substituting back $t = 1 + x^2$ results in

$$\frac{d}{dx}\frac{1}{1+x^2} = -\frac{2x}{(1+x^2)^2}$$

Suppose, however, that we are unaware of this result, and need to evaluate the integral $\int [x/(1+x^2)^2]\,dx$. Since the $(1+x^2)^2$ in the denominator is what makes the integral more difficult, it is not unreasonable to try to simplify things by making the change of variable from x to t, where t is given by $t = 1 + x^2$. The integral then becomes $\int (x/t^2)\,dx$.

Of course the integrand can hardly be left in this form, involving **both** x and t, and the aim is to arrive at an integral of a function of t. We can certainly get rid of the x in the numerator by writing $x = \pm\sqrt{t - 1}$, the sign depending on the sign of x. Although this does not look too hopeful, there is an even greater difficulty — what do we do with 'dx' when to integrate a function of t would presumably require 'dt'? Fortunately, the notation which we use in calculus was established in the first place to make this kind of thing really easy. Just as the chain rule $dy/dx = (dy/dt)(dt/dx)$ is an example of doing what is suggested by the notation, so in integration the 'obvious' step is in fact the correct one. All you need to do is to replace 'dx' in the integrand by '$(dx/dt)\,dt$'.

What is actually happening here is that we are making use of an identity $\int f(x)dx = \int f(x)\,(dx/dt)\,dt$, where in the right-hand integral x has to be substituted as a function of t before carrying out the integration. The left-hand integral, in fact, is $F(x) + c$, where F is an antiderivative of f. What is being asserted in the right-hand integral is that $F(x)$ is also an antiderivative of $f(x)dx/dt$, **provided** everything is expressed as a function of t, and differentiation is with respect to t. That is, the identity says that

$$\frac{d}{dt}F(x) = f(x)\frac{dx}{dt}$$

and with $x = x(t)$ this follows from the chain rule, since

$$\frac{d}{dt}F(x) = \frac{d}{dx}F(x).\frac{dx}{dt} = f(x).\frac{dx}{dt}$$

as required.

There are, of course, further questions which may be raised concerning the rigorous justification of this method, particularly since $t = 1 + x^2$ does not define an injective function from x to t. In fact, a rigorous proof of the validity, under stated conditions, of integration by substitution can be carried out. In the present context, however, we can always finally have recourse to differentiation to verify that a result is justified in any particular instance, and in this positive spirit I hope the reader will forgive my insistence on the method rather than its proof.

To recap, then, we have, with $t = (1 + x^2)$, $\int [x/(1+x^2)^2]\,dx = \int (x/t^2)\,dx$, where we can replace 'dx' by '$(dx/dt)dt$'. There are several ways in which to carry out the next step. We can write

$$\frac{dt}{dx} = \frac{d}{dx}(1 + x^2) = 2x$$

so that

$$\frac{dx}{dt} = \frac{1}{dt/dx} = \frac{1}{2x}$$

The integral then becomes

$$\int \frac{x}{t^2}\frac{dx}{dt}\,dt = \int \frac{x}{t^2}\cdot\frac{1}{2x}\,dt = \int \frac{1}{2t^2}\,dt$$

Or we can write $x = \sqrt{t-1}$, $dx/dt = 1/2\sqrt{t-1}$, giving

$$\int \frac{x}{t^2}\,dx = \int \frac{\sqrt{t-1}}{t^2}\cdot\frac{1}{2\sqrt{t-1}}\,dt = \int \frac{1}{2t^2}\,dt$$

as before, with the same result had we taken $x = -\sqrt{t-1}$.

Or we can write

$$\int \frac{x}{t^2}\,dx = \int \frac{x}{t^2}\frac{dx}{dt}\,dt = \int \frac{1}{t^2}\frac{d}{dt}(\tfrac{1}{2}x^2)\,dt = \int \frac{1}{t^2}\frac{d}{dt}(\tfrac{1}{2}(t-1))\,dt = \int \frac{1}{2t^2}\,dt$$

again.

In any case, the main idea after the replacement of 'dx' by '(dx/dt) dt' is to use the equation $t = 1 + x^2$ to express **everything** in terms of t. However this is carried out, we can now perform the integration with respect to t, which gives

$$\int \frac{x}{(1+x^2)^2}\,dx = \int \frac{1}{2t^2}\,dt = -\frac{1}{2t} + c$$

Substituting back $t = (1 + x^2)$ now yields the final result

$$\int \frac{x}{(1+x^2)^2}\,dx = -\frac{1}{2(1+x^2)} + c$$

which is what we expected following our initial discussion of this example, and may be confirmed by direct differentiation.

The successful use of integration by substitution is highly dependent on an appropriate change of variable. This is often a matter of common sense, experience and luck. The following example illustrates how the use of known identities, particularly for trigonometric or hyperbolic functions, may be an important factor.

✑ *Example 8*

Integration by substitution: we consider the two integrals $\int [1/(4 + x^2)]\,dx$ and $\int [1/\sqrt{x(x+1)}]\,dx$ $(x > 0)$. In the first instance, we may be guided by the trigonometric identity $1 + \tan^2 \theta = \sec^2 \theta$, as well as the result that

$$\frac{d}{d\theta} \tan \theta = \frac{d}{d\theta} \left(\frac{\sin \theta}{\cos \theta} \right) = \frac{\cos \theta \frac{d}{d\theta} \sin \theta - \sin \theta \frac{d}{d\theta} \cos \theta}{\cos^2 \theta}$$

$$= \frac{\cos^2 \theta + \sin^2 \theta}{\cos^2 \theta} = \frac{1}{\cos^2 \theta} = \sec^2 \theta$$

The change of variable $x = \tan \theta$ is not likely to be successful, since then $1/(4 + x^2) = 1/(4 + \tan^2 \theta)$, and little can be done. On the other hand, with a little thought we may want to try $x = 2 \tan \theta$, in which case $1/(4 + x^2) = 1/4(1 + \tan^2 \theta) = 1/4 \sec^2 \theta$. Making this substitution, the integral becomes

$$\int \frac{1}{4 + x^2} \, dx = \int \frac{1}{4 \sec^2 \theta} \frac{dx}{d\theta} \, d\theta = \int \frac{1}{4 \sec^2 \theta} \frac{d}{d\theta} (2 \tan \theta) d\theta$$

$$= \int \frac{1}{4 \sec^2 \theta} . 2 \sec^2 \theta \, d\theta = \int \tfrac{1}{2} d\theta = \tfrac{1}{2} \theta + c$$

Substituting back in terms of x, $x = 2 \tan \theta$ gives $\theta = \tan^{-1}(x/2)$, and we have

$$\int \frac{1}{4 + x^2} \, dx = \tfrac{1}{2} \tan^{-1} \left(\frac{x}{2} \right) + c$$

Our second case, that of the integral $\int [1/\sqrt{x(x + 1)}] \, dx$ is perhaps not quite so clear. The square root in the denominator may, however, suggest the change of variable $x = t^2$. Making this substitution, we now have

$$\int \frac{1}{\sqrt{x(x + 1)}} \, dx = \int \frac{1}{\sqrt{t^2(1 + t^2)}} \frac{dx}{dt} \, dt = \int \frac{1}{t\sqrt{1 + t^2}} . 2t \, dt = \int \frac{2}{\sqrt{1 + t^2}} \, dt$$

where we have taken $t > 0$ and $\sqrt{t^2} = +t$. You may already know an antiderivative for the function $1/\sqrt{1 + t^2}$. If you do not, it turns out after some trial and error that a useful substitution is $t = \sinh \theta$, to make use of the identity $\cosh^2 \theta - \sinh^2 \theta = 1$, satisfied by the hyperbolic cosine and sine functions. With $t = \sinh \theta$, we have

$$\int \frac{2}{\sqrt{1 + t^2}} \, dt = \int \frac{2}{\sqrt{1 + \sinh^2 \theta}} \frac{dt}{d\theta} \, d\theta = \int \frac{2}{\sqrt{\cosh^2 \theta}} \frac{d}{d\theta} (\sinh \theta) d\theta$$

Since $\cosh \theta > 0$ and $d(\sinh \theta)/d\theta = \cosh \theta$, this becomes

$$\int \frac{2}{\sqrt{1 + t^2}} \, dt = \int 2 d\theta = 2\theta + c = 2 \sinh^{-1} t + c$$

Hence our original integral is now given by

$$\int \frac{1}{\sqrt{x(x + 1)}} \, dx = 2 \sinh^{-1} t + c$$

which on substituting back $t = \sqrt{x}$ leads to the final result

$$\int \frac{1}{\sqrt{x(x + 1)}} \, dx = 2 \sinh^{-1} \sqrt{x} + c$$

(The substitution $t = -\sqrt{x}$ leads to the same result.)

Partial fractions

The use of partial fractions as a method of integration depends on the linearity of the operation of integration. In fact, we are already using linearity for integration even in the simplest of examples, say when we write $\int(x^2 + \cos x)\mathrm{d}x = (x^3/3) + \sin x + c$. Here we are using the facts that $x^3/3$ is an antiderivative of x^2 and $\sin x$ an antiderivative of $\cos x$, to assert that the sum of these two antiderivatives will be an antiderivative of $x^2 + \cos x$. In general, if $F(x)$ and $G(x)$ are antiderivatives of $f(x)$ and $g(x)$, respectively, then $AF(x) + BG(x)$ will be an antiderivative of $Af(x) + Bg(x)$, for any constants A and B; to verify this, simply use the linearity of differentiation to show that

$$\frac{\mathrm{d}}{\mathrm{d}x}(AF + BG) = A\frac{\mathrm{d}F}{\mathrm{d}x} + B\frac{\mathrm{d}G}{\mathrm{d}x} = Af + Bg$$

Hence, we can write

$$\int (Af(x) + Bg(x))\mathrm{d}x = A\int f(x)\mathrm{d}x + B\int g(x)\mathrm{d}x$$

Partial fractions in integration are used to integrate **rational** functions, that is functions which can be expressed as a ratio of two polynomials. The method can be used whenever the denominator of the integrand factorizes into the product of two or more (non-constant) polynomials, and the idea is then to express the integrand as a sum of rational functions each of which can be integrated. To explain the method, it is best to start with an example.

⁂ *Example 9*

Integration by partial fractions: consider the indefinite integral $\int [(x+1)/(x^2 - 3x + 2)]\,\mathrm{d}x$. Here the denominator factorizes as $(x-1)(x-2)$, so we are looking at the integral $\int [(x+1)/(x-1)(x-2)]\,\mathrm{d}x$. Notice first of all that $(x+1)/(x-1)(x-2)$ is a rational function having the properties:
(a) the polynomial in the numerator is of lower degree than the polynomial in the denominator;
(b) there are no repeated factors in the denominator.
   The theory of partial fractions tells us in all such cases that our rational function may be written as a **sum** of rational functions each having the properties (a) the polynomial in the numerator is again of lower degree than the polynomial in the denominator, and (b) the respective **denominators** of the individual terms of the sum are the **factors** of the denominator of the integrand.
   In this example, the factors of the denominator of the integrand are $(x-1)$ and $(x-2)$, so our sum will be of two rational functions, one having denominator $(x-1)$ and the other having denominator $(x-2)$. These two denominators being polynomials of degree one, their respective numerators, which must have lower degree, can only be polynomials of degree zero, that is constants; so the theory of partial fractions in this case tells us that we can write the integrand in the form

$$\frac{x+1}{(x-1)(x-2)} = \frac{A}{(x-1)} + \frac{B}{(x-2)} \tag{8.3}$$

where A and B are constants.

Equation (8.3) must hold for all $x \in \mathbb{R}$, except obviously for $x = 1$ and $x = 2$. The next question is, how do we determine the values of the constants A and B? One of the simplest methods is to multiply the equation throughout by the denominator $(x - 1)(x - 2)$ of the left-hand side, to obtain

$$x + 1 = A(x - 2) + B(x - 1) \tag{8.4}$$

A simple argument, based on the continuity of polynomials, shows that equation (8.4), which holds in the first instance for all x **except** $x = 1$ and $x = 2$, holds also when $x = 1$ and $x = 2$, and hence is an identity that is valid for all $x \in \mathbb{R}$.

It is now difficult **not** to be able to determine A and B, there are so many equations at our disposal. We can, for example, substitute any values of x that we like into equation (8.4), to arrive at linear algebraic equations for A and B, which can then be solved, or we can compare the coefficients of x and the constant term on the right and left.

The simplest approach in this case is to substitute, respectively, $x = 1$ and $x = 2$ into equation (8.4): then $x = 1 \Rightarrow 2 = -A \Rightarrow A = -2$; and $x = 2 \Rightarrow 3 = +B \Rightarrow B = 3$.

[Comparison of coefficients of x and the constant term yields

$$\left. \begin{array}{ll} x: & 1 = A + B \\ 1: & 1 = -2A - B \end{array} \right\} \text{ again leading to the solution } A = -2, B = 3]$$

Hence

$$\frac{x + 1}{(x - 1)(x - 2)} = \frac{-2}{(x - 1)} + \frac{3}{(x - 2)}$$

The linearity of integration now allows us to write

$$\int \frac{x + 1}{(x - 1)(x - 2)} dx = -2 \int \frac{1}{(x - 1)} dx + 3 \int \frac{1}{(x - 2)} dx$$

The integrand is undefined at $x = 1$ and at $x = 2$, and there are three separate intervals on which we can look for antiderivatives, namely $-\infty < x < 1$, $1 < x < 2$, and $2 < x < \infty$. In each interval, care has to be taken to write down the correct antiderivative. For example, in the interval $1 < x < 2$ an antiderivative of $1/(x - 1)$ is certainly $\ln(x - 1)$, but an antiderivative of $1/(x - 2)$ is **not** $\ln(x - 2)$, nor is it $-\ln(x - 2)$, neither function being defined for $x < 2$. A function which **is** defined for $x < 2$ is $\ln(2 - x)$. However, many students at this point write down the antiderivative incorrectly as $-\ln(2 - x)$. This is not correct either because, by the chain rule, with $y = -\ln(t)$ and $t = 2 - x$, we have $dy/dx = -(1/t).dt/dx = +1/t = 1/(2 - x)$, which has the wrong sign. There is only one way to be sure, and that is to verify that your antiderivative is in fact defined on the interval in question, and then to **carry out the differentiation**. That way, you are sure to come up with $\ln(2 - x)$ as an antiderivative of $1/(x - 2)$ on the interval $1 < x < 2$, and we have in that case

$$\int \frac{x + 1}{(x - 1)(x - 2)} dx = -2 \ln(x - 1) + 3 \ln(2 - x) + c \quad (1 < x < 2)$$

In fact, a single expression will deal with all three of the intervals in this example, though strictly with a different constant c in each case, and we can write

$$\int \frac{x+1}{(x-1)(x-2)} dx = -2\ln|x-1| + 3\ln|x-2| + c$$

Example 9 is typical of the use of partial fractions in integration. The following example presents a case where the factors are not all linear, and explains how to extend the method a little further.

❀ Example 10

Extension of method of partial fractions: we shall consider three indefinite integrals in this example:

$$I_1 = \int \frac{1}{(x-1)(x^2+1)} dx, \quad I_2 = \int \frac{x+1}{(x-1)^2(x-2)} dx, \quad I_3 = \int \frac{x^2+1}{x^2-1} dx$$

In the first case, the partial fraction of the integrand takes the form $1/(x-1)(x^2+1) = [A/(x-1)] + [(Bx+C)/(x^2+1)]$. Note the numerator $Bx + C$ on the right-hand side, which has to be of lower degree than the denominator $(x^2 + 1)$.

Multiplying the equation throughout by $(x-1)(x^2+1)$ now leads to $1 = A(x^2 + 1) + (Bx + C)(x - 1)$. Now

$$x = 1 \Rightarrow 1 = 2A \Rightarrow A = 1/2$$
$$x = 0 \Rightarrow 1 = A - C \Rightarrow C = -1/2$$

We could now substitute another value of x to determine B, but it may be simpler to equate coefficients of x^2 on each side of the equation, to give

$$x^2: \quad 0 = A + B; \text{ hence } B = -\frac{1}{2}$$

So

$$\int \frac{1}{(x-1)(x^2+1)} dx = \frac{1}{2} \int \frac{1}{(x-1)} dx - \frac{1}{2} \int \frac{x}{(x^2+1)} dx - \frac{1}{2} \int \frac{1}{(x^2+1)} dx$$

There are two intervals involved in this case, namely $-\infty < x < 1$ and $1 < x < \infty$. A common expression will do for both, and we have $I_1 = \frac{1}{2}\ln|x-1| - \frac{1}{4}\ln(x^2+1) - \frac{1}{2}\tan^{-1}x + c$.

To evaluate I_2, note that there is a repeated factor $(x-1)^2$ in the denominator of the integrand, and the usual partial fraction expansion must be adapted somewhat to deal with this situation. The simplest approach is to make use of the previous result (in Example 9) without the repeated factor. In that case, we had $(x+1)/(x-1)(x-2) = [-2/(x-1)] + [3/(x-2)]$, which on multiplying throughout by $1/(x-1)$ gives

$$\frac{x+1}{(x-1)^2(x-2)} = \frac{-2}{(x-1)^2} + \frac{3}{(x-1)(x-2)}$$

The second term on the right can again be expanded in partial fractions, and we end up with

$$\frac{x+1}{(x-1)^2(x-2)} = \frac{-2}{(x-1)^2} - \frac{3}{(x-1)} + \frac{3}{(x-2)}$$

Carrying out the integration, we now have

$$\int \frac{x+1}{(x-1)^2(x-2)} = \frac{2}{(x-1)} - 3\ln|x-1| + 3\ln|x-2| + c$$

The final integration, $I_3 = \int [(x^2+1)/(x^2-1)]\,dx$, falls just outside the general theory which I have outlined, since the numerator in this case is a polynomial of the same degree as the denominator; but it is simple to remedy this by division, and so $(x^2+1)/(x^2-1)$ can be written in the form $(x^2+1)/(x^2-1) = 1 + [2/(x^2-1)]$. Using a partial fraction expansion on the right-hand side now leads to the result

$$I_3 = \int 1\,dx + \int \frac{1}{(x-1)}\,dx - \int \frac{1}{(x+1)}\,dx = x + \ln|x-1| - \ln|x+1| + c$$

EXERCISES ON 8.3

1. Given a function $f(x)$, write down antiderivatives of each of the following functions:

$$2\frac{df}{dx}, \quad x\frac{df}{dx} + f, \quad \frac{x\frac{df}{dx} - f}{x^2}, \quad \frac{f\frac{d^2f}{dx^2} - (\frac{df}{dx})^2}{f^2}, \quad 2\left(\frac{df}{dx}\right)\left(\frac{d^2f}{dx^2}\right) + 2f\frac{df}{dx}, \quad f'(4x)$$

2. Find a partial fraction expansion for the function $1/x(x-1)^3$ which will enable you to evaluate the integral

$$\int \frac{1}{x(x-1)^3\,dx}$$

3. Evaluate $\int |x|\,dx$ (a) for $x > 0$ and (b) for $x < 0$.
4. Use integration by parts to show that $\int e^x \cos x\,dx = e^x \sin x - \int e^x \sin x\,dx$, and integrate by parts again to show that $\int e^x \cos x\,dx = e^x(\sin x + \cos x) - \int e^x \cos x\,dx$. Hence evaluate the integral $\int e^x \cos x\,dx$, and use the same method to evaluate $\int e^{-x} \sin x\,dx$. What other method do you know to evaluate this integral?
5. Integration by parts applied to the integral $\int (1/x)\,((d/dx)\,x)\,dx$ seems to show that $\int (1/x)\,dx = 1 - \int -(1/x)\,dx$, or $\int (1/x)\,dx = 1 + \int (1/x)\,dx$, or $0 = 1$! Explain this apparent inconsistency with reference to the definition of the indefinite integral.
6. Find a change of variable which will help you to evaluate the integral $\int (\ln(x))^3\,dx$, and then use integration by parts to determine the integral.
7. Evaluate the integral $\int [1/x(x^2+1)]\,dx$ (a) by use of partial fractions and (b) by means of the change of variable $x = 1/t$, followed by one further change of variable.

Summary

In this chapter, I have introduced the antiderivative, or indefinite integral. The two terms are not quite identical; one speaks of an antiderivative of a function f — this is a function F for which $F' = f$, whereas the indefinite integral is an expression, $\int f(x)dx = F(x) + c$, for all possible antiderivatives. Antidifferentiation is the inverse operation to differentiation, and as such your understanding of antidifferentiation will be strongly dependent on your previous mastery of the fundamentals of differentiation. Apart from inspired guesswork, which is sometimes given the more respectable title of 'integration by inspection', I have given you three methods of carrying out an indefinite integral. These are the methods of integration by parts, substitution and partial fractions. Each one of these has its counterpart in the theory of differentiation, and certainly you should make sure you have fully understood the corresponding ideas in differentiation before taking the further step.

As a useful aid in integration, I shall list below a number of standard indefinite integrals. This is essentially the inverse of the list of derivatives which I gave you at the end of Chapter 5, but supplemented by some further results of Chapter 7, involving exponential, logarithmic and related functions.

- $\int (Af + Bg)dx = A \int f dx + B \int g dx$; A, B constants (LINEARITY)

- $\int f \frac{dg}{dx} dx = fg - \int g \frac{df}{dx} dx$ (INTEGRATION BY PARTS)

- $\int f dx = \int f(x(t)) \frac{dx}{dt} dt$ (CHANGE OF VARIABLE)

- $\int x^n dx = \frac{x^{n+1}}{n+1} + c$, n an integer $\neq -1$

- $\int \frac{1}{x} dx = \ln(|x|) + c$

- $\int (\sum_0^\infty a_n x^n)dx = \sum_0^\infty \frac{a_n x^{n+1}}{(n+1)} + c$; $|x| < R =$ radius of convergence of the integrated series (INTEGRATION OF POWER SERIES)

- $\int x^k dx = \frac{x^{k+1}}{k+1} + c$, for $x > 0$, k a real number $\neq -1$

- $\int \sin(ax)dx = -\frac{1}{a}\cos(ax) + c$ (a constant $\neq 0$)

- $\int \cos(ax)dx = \frac{1}{a}\sin(ax) + c$

- $\int e^{ax} dx = \frac{1}{a}e^{ax} + c$

- $\int \sinh(ax)dx = \frac{1}{a}\cosh(ax) + c$

- $\int \cosh(ax)dx = \frac{1}{a}\sinh(ax) + c$

- $\int \frac{1}{1+x^2} dx = \tan^{-1} x + c$

- $\int \frac{1}{\sqrt{1-x^2}} dx = \sin^{-1} x + c$

- $\int \frac{1}{\sqrt{x^2-1}} dx = \cosh^{-1} x + c$

- $\int \frac{1}{\sqrt{x^2+1}} dx = \sinh^{-1} x + c$

FURTHER EXERCISES

1. Find antiderivatives for the functions $\cos 4x$, $x \exp(2x^2)$, $\cos^5 x \,(= \cos x \,(1 - \sin^2 x)^2)$, and $\cosh^4 x$, in the last case by expressing the function in terms of the exponential.

2. By making appropriate changes of variable, evaluate the indefinite integrals $\int \exp \sqrt{x} dx$, $\int \frac{x^3}{(1+x^4)^2} dx$, $\int \frac{(\ln(x))^2}{x} dx$, $\int \frac{x^4}{(1+x)} dx$, $\int x^2 \ln(x) dx$ and $\int \sin^{-1} x dx$.

3. Use integration by parts to evaluate the integrals $\int x^2 \cos 2x \, dx$, $\int x^4 \ln(x) dx$ and $\int \tan^{-1} x dx$.
   Use integration by parts to show that

$$\int \sqrt{1 + x^2} dx = x\sqrt{1 + x^2} - \int \frac{x^2}{\sqrt{1 + x^2}} dx$$

By writing $\int (x^2/\sqrt{1 + x^2}) \, dx = \int \{\sqrt{1 + x^2} - (1/\sqrt{1 + x^2})\} dx$, evaluate the integral $\int \sqrt{1 + x^2} dx$.

4. Use the change of variable $x = \sinh \theta$, together with identities for hyperbolic functions listed at the end of Chapter 7, to evaluate the integral $\int \sqrt{1 + x^2} dx$.

5. Verify that

$$\frac{d}{dx} \tanh^{-1} x = \frac{1}{1 - x^2} \quad (-1 < x < 1)$$

Use partial fractions to find an antiderivative of $1/(1 - x^2)$ in this interval, and hence write down an identity for $\tanh^{-1} x$ in terms of the natural logarithmic function.

6. The speed v of a particle falling under gravity is known to satisfy the equation $dv/dt = c^2 - k^2 v^2$, where c and k are positive constants. Assuming $v < c/k$, and that the particle is at rest at time $t = 0$, use the result of exercise 5 to express v as a function of t involving the hyperbolic tangent function. Hence show that the particle approaches a limiting speed c/k as t tends to infinity.

7. The distance travelled by an object after time t is given by $s = s(t)$. If the speed of the object is given by $ds/dt = 1 + s^2$, by considering $(d/dt) \tan^{-1} s$ show that $s(t) = \tan t$.

8. The height $z(t)$ of a particle moving under gravity is known to satisfy the equation $d^2z/dt^2 = -g$. By considering $(d/dt) ((dz/dt)^2 + 2gz)$, show that if the velocity dz/dt is u_0 when the particle is at height $z = 0$, then the velocity at general time t satisfies the equation $(dz/dt)^2 = u_0^2 - 2gz$.

9. Use the substitution $x = \tan \theta$, together with the identity $\cos^2 \theta = (1 + \cos 2\theta)/2$, to evaluate the integral $\int \frac{1}{(1+x^2)^2} dx$.

10. Use partial fractions to evaluate the integrals (on some appropriate interval)

$$\int \frac{x}{(x^2 - 1)} dx, \quad \int \frac{1}{x^2(x^2 + 1)} dx, \quad \int \frac{(x - 1)}{(x + 1)(x - 2)^2} dx \text{ and } \int \frac{x^3}{x^2 - 1} dx$$

11. Use partial fractions to evaluate the integral $\int \frac{1}{x(x^2+1)^2} dx$.

9 • The Definite Integral

An antiderivative is a function, an indefinite integral is a set of functions; but a definite integral is a number. How is this number defined? Given a function $f(x)$ on a finite interval $a \leq x \leq b$, the definite integral of f across the interval, written $\int_a^b f(x)\mathrm{d}x$, may be thought of as the area between the graph $y = f(x)$ and the x-axis, for values of x in the given interval. Areas above the x-axis, where $f(x) > 0$, count as positive, and areas below the x-axis count as negative. If t is the input variable, then the same definite integral will be denoted by $\int_a^b f(t)\mathrm{d}t$. Thus $\int_a^b f(t)\mathrm{d}t = \int_a^b f(x)\mathrm{d}x$.

In Fig 9.1, $\int_a^b f(x)\mathrm{d}x$ is the shaded area indicated, whereas in Fig 9.2 $\int_a^b f(x)\mathrm{d}x$ is the shaded area above the x-axis, marked with a plus, minus the shaded area below the x-axis, marked with a minus.

It will be anticipated from the notation that $\int_a^b f(x)\mathrm{d}x$ has something to do with antiderivatives and the indefinite integral, though it may not be so obvious *ab initio* how these different types of integral are related. In this chapter, I shall set out the properties of the definite integral and show how it is to be determined. We shall see too that the definite integral is not only an area, but also has many other interpretations.

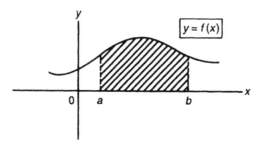

Fig 9.1 A definite integral.

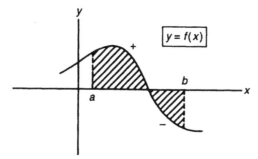

Fig 9.2 A definite integral with signs.

9.1 The definite and indefinite integral

Although the integral as area gives an intuitive idea of the definite integral, this by no means amounts to a formal definition, and many questions remain to be answered. We are all familiar with areas of simple geometrical figures, but precisely what is meant by the area between a general curve and the x-axis? We know that there are functions which do not have a derivative. Presumably too there may be functions for which $\int_a^b f(x)\mathrm{d}x$ just does not make sense. How can we be sure which functions are integrable across an interval, and which are not?

To give final answers to these questions would require an excursion into real analysis which is beyond the scope of this book. Nevertheless, I think it is helpful to have at least a cursory knowledge of the directions that a more formal treatment can take. I say directions and not direction, because there are several possible ways of defining the definite integral, each appropriate in its own context. The first of these is the so-called Riemann integral. To define the Riemann integral $\int_a^b f(x)\mathrm{d}x$, you divide the interval $a \le x \le b$ into subintervals $a \le x < x_1$, $x_1 \le x < x_2$, $x_2 \le x < x_3$, ..., $x_{n-1} \le x \le b$. For each subinterval $x_k \le x < x_{k+1}$, define an element of area to be $f(x).(x_{k+1} - x_k)$, where x is some point belonging to the subinterval and $x_{k+1} - x_k$ is the length of the subinterval. The Riemann integral is then defined to be the limiting sum over all these subintervals of these elements of area, as the maximum length of a subinterval tends to zero and the number of subintervals tends to infinity. Of course, for this definition to work, the limit must both exist and be independent of the precise details of the way in which the interval $a \le x \le b$ is subdivided.

In fact, the Riemann integral can be shown to exist whenever the function $f(x)$ is both defined and continuous at every $x \in (a, b)$, as well as right continuous at $x = a$ and left continuous at $x = b$. Moreover, as you would expect, the Riemann integral agrees with our intuitive idea of area, wherever a comparison can be made.

A second way of defining the definite integral is the so-called Lebesgue integral. The Lebesgue integral, the analysis of which requires a knowledge of measure theory, is more powerful in that there is a much wider class of functions which can be integrated than is the case with the Riemann integral.

Yet other ways of defining the definite integral are possible, and there is no one 'best' definition. Fortunately, it is not necessary for us to enter into all the details of these different definitions. For our purposes, it will be sufficient to know that $\int_a^b f(x)\,\mathrm{d}x$ can at least be defined for any function f continuous on the interval $[a, b]$, and that whatever definition of integral is adopted, the following properties will hold, which agree with our intuitive understanding of area.

Properties of the definite integral

(1) $\int_a^b f(x)\mathrm{d}x = -\int_b^a f(x)\mathrm{d}x$. [This defines $\int_a^b f(x)\mathrm{d}x$ even if $b \le a$, and if $b = a$ has the consequence $\int_a^a f(x)\mathrm{d}x = 0$. You can change the value taken by a function f at just a single point without changing the total integral.]

(2) $\int_a^b (Af(x) + Bg(x))\mathrm{d}x = A\int_a^b f(x)\mathrm{d}x + B\int_a^b g(x)\mathrm{d}x$, A, B constants (LINEARITY).

(3) $\int_a^b f(x)\mathrm{d}x = \int_a^c f(x)\mathrm{d}x + \int_c^b f(x)\mathrm{d}x$.

(4) $f(x) \geq g(x) \Rightarrow \int_a^b f(x)\mathrm{d}x \geq \int_a^b g(x)\mathrm{d}x$ for $b > a$.

(5) $\int_a^b 1\mathrm{d}x = (b - a)$.

I hope that you will accept that (1)–(5) above agree with your intuitions about area. Property (1) is more of a definition than anything else. Property (2) describes how areas behave under multiplication by real numbers and addition; if you double the height of the graph you will double the area, if you put one area on top of another you must add their respective areas. Property (3) is also about adding areas, this time describing the addition of areas that are placed side to side. Property (4) says that if you increase the height then you will increase the area. Finally (5) gives the area of a rectangle of unit height.

Natural and obvious though (1)–(5) appear to be, these properties in fact contain enough information, in principle, to define uniquely the definite integral of a function continuous on a closed interval. In problem 1 of Exercises on 9.1, for example, these properties alone are used to show that $\int_0^1 x\mathrm{d}x = \frac{1}{2}$. Given that (1)–(5) are acceptable starting points for our discussion of the definite integral, and that they contain within them, though we shall not make this explicit, a **definition** of the definite integral, we can hardly go wrong in relying on these properties to tell us how to **evaluate** the definite integral. The key to this is in the following theorem, which provides a link between definite integrals on the one hand and indefinite integrals on the other.

● *Theorem I (Fundamental Theorem of Calculus)* ————

Suppose $f(x)$ is defined and continuous at all x on a closed interval $a \leq x \leq b$. Then

$$\frac{\mathrm{d}}{\mathrm{d}x} \int_a^x f(t)\mathrm{d}t = f(x) \ (a \leq x \leq b)$$

That is, the function $\int_a^x f(t)\mathrm{d}t$ is an antiderivative of $f(x)$ on the prescribed interval.

PROOF

Before following through the proof, we should be as clear as possible about notation. $\int_a^b f(x)\mathrm{d}x$ is a number, depending not only on the function f but also on the interval $[a, b]$; if you change the interval then you will also change the integral. The integral $\int_a^b f(t)\mathrm{d}t$ is also a number, being the same number as $\int_a^b f(x)\mathrm{d}x$; here t rather than x is the input variable, but the function f and the interval $[a, b]$ are still the same. When we write down an integral such as $\int_a^x f(t)\mathrm{d}t$ (or $\int_a^x f(\theta)\mathrm{d}\theta$, or $\int_a^x f(s)\mathrm{d}s$, all these integrals being the same), we are allowing the upper endpoint of the interval of integration to vary. This x can be thought of as an input variable. Thus $\int_a^x f(t)\mathrm{d}t$ depends on the value of x, and defines a **function** of x. It is this function which is differentiated in the statement of the theorem. I prefer **not** to use the notation $\int_a^x f(x)\mathrm{d}x$, since it is generally best to use different symbols for the limits of integration and for the input variable.

Now, to carry out the differentiation let $I(x) = \int_a^x f(t)\mathrm{d}t$ and consider the limit $\lim_{h\to 0}[I(x+h) - I(h)]/h$, which on making use of property (3) above results in

$$\lim_{h\to 0}\frac{\int_a^{x+h} f(t)\mathrm{d}t - \int_a^x f(t)\mathrm{d}t}{h} = \lim_{h\to 0}\frac{1}{h}\int_x^{x+h} f(t)\mathrm{d}t$$

Consider the case of h small and positive. The value of the integral on the right-hand side depends only on the interval $[x, x+h]$ and the behaviour of the function $f(t)$ on this interval. Since f is continuous at x, we also know that the **variation** $V(h)$ of f over this interval must go to zero as h approaches zero. In fact, on the given interval, $f(t)$ must lie between $f(x) - V(h)$ and $f(x) + V(h)$; but in that case, with $f(x) - V(h) \le f(t) \le f(x) + V(h)$, property (4) above allows us to compare integrals, with the result that

$$\frac{1}{h}\int_x^{x+h}(f(x) - V(h))\mathrm{d}t \le \frac{1}{h}\int_x^{x+h} f(t)\mathrm{d}t \le \frac{1}{h}\int_x^{x+h}(f(x) + V(h))\mathrm{d}t$$

Since neither $f(x) - V(h)$ nor $f(x) + V(h)$ depends on t, the two outer integrals in the inequality may be carried out immediately, and we have

$$f(x) - V(h) \le \frac{1}{h}\int_x^{x+h} f(t)\mathrm{d}t \le f(x) + V(h)$$

Hence $(1/h)\int_x^{x+h} f(t)\mathrm{d}t$ is sandwiched between two expressions, both of which converge to $f(x)$ in the limit as $h \to 0+$, since we know from the continuity of f that $\lim_{h\to 0} V(h) = 0$.

A similar argument applies to the limit as $h \to 0$ through negative values, and we can conclude that

$$\frac{\mathrm{d}}{\mathrm{d}x}\int_a^x f(t)\mathrm{d}t = f(x)$$

as required.

Theorem 1 has two important implications. The first of these is that the theorem confirms the result, stated earlier, that a continuous function on an interval will have an antiderivative on that interval; $\int_a^x f(t)\mathrm{d}t$ is an antiderivative of $f(x)$. Secondly, the theorem provides a method of evaluating the integral $\int_a^b f(x)\mathrm{d}x$. Let $F(x)$ be **any** antiderivative of $f(x)$ on the given interval. Since two anti derivatives differ by a constant, we then have $F(x) = \int_a^x f(t)\mathrm{d}t + c$, where c is a constant. Hence $F(b) - F(a) = \{(\int_a^b f(t)\mathrm{d}t + c) - c\} = \int_a^b f(t)\mathrm{d}t = \int_a^b f(x)\mathrm{d}x$, so we have the following easily applicable rule for evaluating the integral $\int_a^b f(x)\mathrm{d}x$, provided f is both defined and continuous at each point of the interval (in fact at $x = a$ it is sufficient for f to be right continuous, and left continuous at b): find an antiderivative $F(x)$ of $f(x)$ on the interval. Then the integral is given by

$$\int_a^b f(x)\mathrm{d}x = \int_a^b \frac{\mathrm{d}}{\mathrm{d}x}F(x)\mathrm{d}x = F(b) - F(a) \tag{9.1}$$

Equation (9.1) is the fundamental result for the evaluation of definite integrals. We can even use this result to evaluate definite integrals of discontinuous functions, since if, say, $x = x_0$ is the only point of discontinuity of f in the interval, it may still be possible to evaluate $\int_a^{x_0} f(x)\mathrm{d}x$ and $\int_{x_0}^b f(x)\mathrm{d}x$ separately, and then use property (3) to write $\int_a^b f(x)\mathrm{d}x = \int_a^{x_0} f(x)\mathrm{d}x + \int_{x_0}^b f(x)\mathrm{d}x$.

A word of warning! In using equation (9.1) to evaluate $\int_a^b f(x)\mathrm{d}x$, it is **essential** that we have $\mathrm{d}(F(x))/\mathrm{d}x = f(x)$ for **all** x in the interval; that is, $F(x)$ must be a true antiderivative on the interval. It would **not** be correct to write $\int_{-1}^1 (1/x^2)\mathrm{d}x = \int_{-1}^1 (\mathrm{d}/\mathrm{d}x)(-1/x)\mathrm{d}x = -1 - (+1) = -2$, since it is **not** true that $(\mathrm{d}/\mathrm{d}x)(-1/x) = 1/x^2$ at $x = 0$. The fact that the integrand is positive on the interval in question should convince you that the integral could not possibly be -2; this integral is in fact undefined.

The following examples will illustrate the application of these ideas. We shall use the common notation of writing $[F(x)]_a^b$ or $[F(x)]_{x=a}^{x=b}$ for $F(b) - F(a)$.

❋ Example 1

Definite integrals: consider first the integral $\int_0^1 x^2 e^x \mathrm{d}x$. An antiderivative of $x^2 e^x$ which may be obtained, for example, by the method of integration by parts, is $(x^2 - 2x + 2)e^x$. Hence $\int_0^1 x^2 e^x \mathrm{d}x = [(x^2 - 2x + 2)e^x]_0^1 = (e - 2)$.

In the case of the integral $\int_2^3 [1/(x-1)(x^2+1)]\mathrm{d}x$, note that the function $1/(x-1)(x^2+1)$, which is undefined at $x = 1$, has an antiderivative for $-\infty < x < 1$, and an antiderivative for $1 < x < \infty$. Here the interval of integration is $[2,3]$, which is contained entirely to the **right** of $x = 1$, so only an antiderivative for $1 < x < \infty$ is relevant. In fact, an antiderivative has already been found in Example 10 of Section 8.3. For $x > 1$, we have

$$\int \frac{1}{(x-1)(x^2+1)}\mathrm{d}x = \frac{1}{2}\ln(x-1) - \frac{1}{4}\ln(x^2+1) - \frac{1}{2}\tan^{-1}x + c$$

Hence we have

$$\int_2^3 \frac{1}{(x-1)(x^2+1)}\mathrm{d}x = \left[\frac{1}{2}\ln(x-1) - \frac{1}{4}\ln(x^2+1) - \frac{1}{2}\tan^{-1}x\right]_{x=2}^{x=3}$$
$$= \frac{1}{2}\ln(2) - \frac{1}{4}\ln(10) - \frac{1}{2}\tan^{-1}3$$
$$- \left\{\frac{1}{2}\ln(1) - \frac{1}{4}\ln(5) - \frac{1}{2}\tan^{-1}2\right\}$$

which simplifies somewhat to

$$\frac{1}{2}\ln(2) - \frac{1}{4}\ln\left(\frac{10}{5}\right) + \frac{1}{2}(\tan^{-1}2 - \tan^{-1}3) = \frac{1}{4}\ln(2) + \frac{1}{2}(\tan^{-1}2 - \tan^{-1}3)$$

❋ Example 2

Integration by substitution: as an example of the evaluation of a definite integral by change of variable, let $I = \int_0^1 [x/(1+x^2)^2]\mathrm{d}x$. We have already obtained an antiderivative for the function $x/(1+x^2)^2$ by the method of change of variable in Section 8.3. In this instance, however, I shall repeat the calculation to show how the choice of proper limits of integration may be incorporated into the evaluation of the definite integral.

The change of variable $t = 1 + x^2$ leads to

$$I = \int_{x=0}^{x=1} \frac{x}{(1+x^2)^2} dx = \int_{x=0}^{x=1} \frac{x}{t^2} \frac{dx}{dt} dt$$

which with $dx/dt = 1/(dt/dx) = 1/2x$ results in $I = \int_{x=0}^{x=1}(1/2t^2)dt$. Notice that in this and similar examples, where both x and t are involved, it is **essential** to indicate explicitly for all limits of integration whether they refer to x or to t. This draws attention to the need to obtain the t-limits in the last expression for I. We can do this by observing, with $t = 1 + x^2$, that as x increases from 0 to 1, so t increases from 1 to 2. Hence the limits of integration with respect to t are $t = 1$ and $t = 2$, and we have finally

$$\int_0^1 \frac{x}{(1+x^2)^2} dx = \int_{t=1}^{t=2} \frac{1}{2t^2} dt = \left[-\frac{1}{2t} \right]_{t=1}^{t=2} = -\frac{1}{4} - \left(-\frac{1}{2} \right) = \frac{1}{4}$$

Example 3

As an illustration of the use of property (3) in the evaluation of integrals, consider the two integrals $I_1 = \int_0^2 |x(x-1)|dx$ and $I_2 = \int_0^2 g(t)dt$, where the function g is defined by

$$g(t) = 1 - t, \quad 0 \leq t < 1$$
$$= 1, \quad 1 \leq t \leq 2$$

In the case of I_1, note that x has a positive sign across the interval of integration $[0,2]$, so that we can replace $|x|$ by x on this interval; remember that **only** the values taken by a function on the interval of integration will affect the final integral. On the other hand, $x - 1$ changes sign at $x = 1$; in fact $|x - 1|$ is $x - 1$ for $x \geq 1$ but $1 - x$ for $x < 1$. Although $|x(x-1)|$ is a continuous function, and therefore has an antiderivative on $[0,2]$, rather than determining this antiderivative, it is much easier to subdivide $[0,2]$ into two subintervals $[0,1]$ and $[1,2]$, and write

$$\int_0^2 |x(x-1)|dx = \int_0^1 |x(x-1)|dx + \int_1^2 |x(x-1)|dx$$
$$= \int_0^1 x(1-x)dx + \int_1^2 x(x-1)dx$$
$$= \left[\frac{x^2}{2} - \frac{x^3}{3} \right]_0^1 + \left[\frac{x^3}{3} - \frac{x^2}{2} \right]_1^2 = \frac{1}{6} + \frac{5}{6} = 1$$

The integrand of I_2 is discontinuous at $t = 1$. Here there is **no** antiderivative on the interval $[0,2]$ (see problem 6 on page 113). In this case we must subdivide the interval of integration. This gives

$$I_2 = \int_0^1 (1 - t)dt + \int_1^2 1 dt = \left[t - \frac{t^2}{2} \right]_0^1 + [t]_1^2 = \frac{3}{2}$$

1. A function f is defined by $f(x) = x$ ($0 \le x \le 1$). Functions g and h are defined respectively on the interval $0 \le x \le 1$ by

$$g(x) = 0 \quad \left(0 \le x \le \frac{1}{N}\right) \qquad\qquad h(x) = \frac{1}{N} \quad \left(0 \le x \le \frac{1}{N}\right)$$

$$= \frac{1}{N} \quad \left(\frac{1}{N} < x \le \frac{2}{N}\right) \qquad = \frac{2}{N} \quad \left(\frac{1}{N} < x \le \frac{2}{N}\right)$$

$$= \frac{2}{N} \quad \left(\frac{2}{N} < x \le \frac{3}{N}\right) \qquad = \frac{3}{N} \quad \left(\frac{2}{N} < x \le \frac{3}{N}\right)$$

$$\vdots \qquad\qquad\qquad\qquad \vdots$$

$$= \frac{N-1}{N} \quad \left(\frac{N-1}{N} < x \le 1\right) \qquad = 1 \quad \left(\frac{N-1}{N} < x \le 1\right)$$

where N is a positive integer. Sketch with the same axes the graphs of the functions $f(x)$, $g(x)$ and $h(x)$, and show that $g(x) \le f(x) \le h(x)$. Use properties (2)–(5) of the definite integral, listed at the start of Section 9.1, to evaluate $\int_0^1 g(x)\mathrm{d}x$ and $\int_0^1 h(x)\mathrm{d}x$, and to show that $\int_0^1 x\mathrm{d}x$ lies between $(N-1)/2N$ and $(N+1)/2N$. By taking the limit $N \to \infty$, deduce that $\int_0^1 x\mathrm{d}x = \frac{1}{2}$.

2. It has been stated that the definite integral $\int_a^b f(x)\mathrm{d}x$ may be evaluated as $F(b) - F(a)$, where F is an antiderivative of f on the interval $a \le x \le b$. Verify that the integral obtained in this way is independent of the particular choice of antiderivative of f. Verify in addition that this method of evaluation of definite integrals is consistent with all of the properties (1)–(5) of definite integrals listed in Section 9.1.

3. Evaluate the integral $\int_{-1}^1 \frac{|t|}{t+2}\mathrm{d}t$.

9.2 Uses of the definite integral

The fundamental theorem of calculus provides us with an explicit antiderivative of any given function f, at least if that function is continuous. Since

$$\frac{\mathrm{d}}{\mathrm{d}x} \int_a^x f(t)\mathrm{d}t = f(x)$$

we know that $\int_a^x f(t)\mathrm{d}t$ is an antiderivative of $f(x)$, and that any antiderivative of f is given by $\int f(x)\mathrm{d}x = \int_a^x f(t)\mathrm{d}t + c$, where c is constant.

The value of the constant a in this expression is to some extent arbitrary, though its choice has to take into account the domain of the function f, and the interval on which an antiderivative is to be defined. For example, $\int_1^x \ln(t)\mathrm{d}t$ is an antiderivative of the function $\ln(x)$ for $x > 0$, but $\int_{-1}^x \ln(t)\mathrm{d}t$ is not, because the lower limit of integration -1 falls outside the given interval $(0, \infty)$, and this integral is not even defined for positive x. However, $\int_c^x \ln(t)\mathrm{d}t$ is an antiderivative of $\ln(x)$ for $x > 0$, for any **positive** constant c. Avoid the common error of writing $(\mathrm{d}/\mathrm{d}x) \int_c^x \ln(t)\mathrm{d}t = \ln(x) - \ln(c)$. This is not true! The point is that, if F is an antiderivative of f, then $dF/dx = f$, and

$$\frac{d}{dx}\int_c^x f(t)dt = \frac{d}{dx}[F(t)]_{t=c}^{t=x} = \frac{d}{dx}\{F(x) - F(c)\} = \frac{d}{dx}F(x) = f(x)$$

There is **no** contribution to the derivative from the term $F(c)$, because $F(c)$ is just a constant, independent of x, so that

$$\frac{d}{dx}\{F(x) - F(c)\} = \frac{dF(x)}{dx} - \frac{dF(c)}{dx} = \frac{dF(x)}{dx} = f(x)$$

as stated; so $(d/dx)\int_c^x \ln(t)dt$ is $\ln(x)$ **not** $\ln(x) - \ln(c)$.

Often in writing down an antiderivative $\int_c^x f(t)dt$ of a given function $f(x)$, a particular choice of the lower limit c may be the most convenient. This will often be so, for example if one has further information about the value of the function at some specific point, say $x = x_0$. In that case, the choice of x_0 as a lower limit of integration may well be appropriate.

⊕ *Example 4*

Suppose a particle starts at the point $x = 1$ at time $t = 0$, and moves subsequently along the x-axis, with velocity $\dot{x}$ given as a function of time by $\dot{x}(t) = t(t+1)$. We have then, two pieces of information. The first may be summarized by the equation

$$\frac{d}{dt}x(t) = t(t+1) \text{ for } t > 0$$

and tells us that $x(t)$ must be an antiderivative of $t(t+1)$. We can then write $x(t) = \int_c^t s(s+1)ds + \text{const}$; as I said earlier, I prefer to use a different symbol s to denote the integration variable from the t which we have as an upper limit of integration.

The second piece of information is that the particle starts at the point $x = 1$ when $t = 0$. This can be summarized in the statement that $x(0) = 1$, or that $x(t) = 1$ when $t = 0$. Since $t = 0$ is special in this instance, it seems sensible to take $c = 0$ as the lower limit of integration in our integral. Thus, $x(t) = \int_0^t s(s+1)ds + c$ gives us the most general expression for x as a function of t. We can now use our further piece of information $x(0) = 1$, often referred to as an initial condition, and substituting $t = 0$ in $x(t)$ gives

$$x(0) = 1 = \int_0^0 s(s+1)ds + c = c$$

where we have used property (1) of definite integrals from Section 9.1.

With $c = 1$, we have

$$x(t) = 1 + \int_0^t s(s+1)ds$$

where the definite integral can easily be evaluated to give, finally

$$x(t) = 1 + \left[\frac{s^3}{3} + \frac{s^2}{2}\right]_{s=0}^{s=t} = 1 + \frac{t^2}{2} + \frac{t^3}{3}$$

Example 4 is typical of the use of the fundamental theorem of calculus, coupled with an initial condition, to obtain an explicit antiderivative. The following example shows the same idea used to great effect to derive some very precise bounds for the exponential function.

⊛ *Example 5*

The function $y(x) = \exp(-x)$ satisfies the equation

$$\frac{\mathrm{d}}{\mathrm{d}x}y(x) = -y(x)$$

Using only this, together with the initial condition $y(0) = 1$, as well as the fact that $y(x)$ is a positive function, we shall be able to determine explicit upper and lower bounds for the exponential function. We take $x > 0$. From the property $\mathrm{d}y/\mathrm{d}x = -y$ we can say immediately that $y(x) = c - \int_0^x y(t)\mathrm{d}t$. However, this equation cannot yet be used to determine $y(x)$, since unfortunately to carry out the integration on the right-hand side would require a knowledge of the function $y(t)$ in the first place!

Using the initial condition $y(0) = 1$ allows us to set $c = 1$ and write

$$y(x) = 1 - \int_0^x y(t)\mathrm{d}t \tag{9.2}$$

Equation (9.2) is called an 'integral equation', in fact a particular kind of integral equation, called a Volterra-type integral equation. I shall not pursue much further here this quite advanced branch of mathematics, except to say that solutions of integral equations such as equation (9.2) are often obtained by the method of iteration. An iterative technique, used to solve equation (9.2), would start with an initial approximation to the solution, say $y = y_0(x)$, and then define a sequence of successive approximations $y = y_1(x)$, $y_2(x)$, $y_3(x)$, ..., given recursively by the equation

$$y_{n+1}(x) = 1 - \int_0^x y_n(t)\mathrm{d}t \tag{9.3}$$

It is then possible to show that, in the limit as n tends to infinity, the sequence of iterative approximations converges to the exact solution; in this case we have $\lim_{n\to\infty} y_n(x) = \exp(-x)$. We shall look at some more examples of iteration in Chapter 10.

Without actually carrying out an iterative solution in this instance, we shall use related ideas to derive a sequence of inequalities for the function $\exp(-x)$ with $x > 0$. Knowing that $y(t)$ is positive, we have $\int_0^x y(t)\mathrm{d}t \geq 0$ for $x \geq 0$, so that equation (9.2) implies $y(x) \leq 1$ (for $x \geq 0$). However, we can now use $y(t) \leq 1$ in the integral to deduce that $\int_0^x y(t)\mathrm{d}t \leq \int_0^x 1\mathrm{d}t = x$. Hence equation (9.2) now implies the further inequality

$$y(x) \geq 1 - x \quad \text{(for } x \geq 0)$$

Now use $y(t) \geq 1 - t$ for $t \geq 0$ to estimate the integral again, giving $\int_0^x y(t)\mathrm{d}t \geq \int_0^x (1 - t)\mathrm{d}t = x - (x^2/2)$. Hence from equation (9.2) we now have

$$y(x) \le 1 - x + \frac{x^2}{2}$$

The next step is to use $y(t) \le 1 - t + (t^2/2)$ to show that $\int_0^x y(t)\mathrm{d}t \le x - (x^2/2) + (x^3/3.2)$. This in turn, from equation (9.2), tells us that

$$y(x) \ge 1 - x + \frac{x^2}{2!} - \frac{x^3}{3!}$$

I hope you can get the picture of what is happening, and see how to extend the sequence of inequalities. For any odd integer value of n, we are led to the following inequalities satisfied by the solution $y(x) = \exp(-x)$ of equation (9.2):

$$1 - x + \frac{x^2}{2!} - \frac{x^3}{3!} \dots - \frac{x^n}{n!} \le \exp(-x) \le 1 - x + \frac{x^2}{2!} - \frac{x^3}{3!} \dots + \frac{x^{n+1}}{(n+1)!}$$

These bounds on the exponential function $\exp(-x)$ for $x > 0$ become more and more precise as the value of n is increased. I hope the terms to the left and right are instantly recognizable as terms of the Taylor expansion for $\exp(-x)$. In fact, the inequalities, which may be used to verify the Taylor expansion, imply that for $x > 0$ the function $\exp(-x)$ lies between two consecutive finite sums of terms of the series. You may wish to use these bounds, for example, to obtain very precise estimates for the value of $1/e$. We not only obtain the value of $1/e$ to a high degree of accuracy, but determine at the same time a range of values in which we can be sure that $1/e$ lies.

In particle dynamics, a definite integral will often have a particular physical interpretation. Integrate the **speed** of a particle with respect to time from $t = t_1$ to $t = t_2$, and you have $\int_{t_1}^{t_2} (\mathrm{d}s/\mathrm{d}t)\, \mathrm{d}t = s(t_2) - s(t_1)$, which is the distance travelled by the particle during the given time interval. Integrate the **velocity**, and you obtain the difference between the coordinates of the particle at the two instants of time. This is the relative displacement, over the time interval. The integral of the **acceleration** is $\int_{t_1}^{t_2} (\mathrm{d}v/\mathrm{d}t)\, \mathrm{d}t = v(t_2) - v(t_1)$, which is the change in velocity over the time interval.

The **rate of working** of a force f acting on a particle in classical mechanics is defined to be $f.\mathrm{d}x/\mathrm{d}t$, and if this is integrated with respect to time you obtain

$$\int_{t_1}^{t_2} f.\frac{\mathrm{d}x}{\mathrm{d}t}\mathrm{d}t = \int_{t_1}^{t_2} m\frac{\mathrm{d}^2x}{\mathrm{d}t^2}\frac{\mathrm{d}x}{\mathrm{d}t}\mathrm{d}t$$

$$= \int_{t_1}^{t_2} \frac{\mathrm{d}}{\mathrm{d}t}\left(\frac{1}{2}m\left(\frac{\mathrm{d}x}{\mathrm{d}t}\right)^2\right)\mathrm{d}t$$

$$= T(t_2) - T(t_1)$$

which is the change in kinetic energy $T = \frac{1}{2}mv^2$ between the two instants of time.

Many further examples of this kind could be given, extending far beyond the boundaries of classical mechanics. Rather than adding to the list, however, I shall give a rather different application of the definite integral, and that is as the integral of a **density**.

A typical example among many is to do with random quantities in probability theory which can assume a continuous range of values. You may be surprised to learn that I wrote this sitting at a desk in Spain. It was rather warm, and there was no air-conditioning in the building. I was interested to know what the temperature would have reached at 3.00 p.m. that afternoon. No one could be sure of the answer, as there is a degree (no pun intended) of uncertainty, but if I were to consult a meteorologist I **might** be able to obtain the information in the form of a function $p(t)$ expressing the probability that the temperature at that time would not exceed a given value t. Thus, for example, $p(t_1) = \text{prob}\{\text{temp} \leq t_1\}$, and $p(t_2) = \text{prob}\{\text{temp} \leq t_2\}$, so that the probability of the temperature at 3.00 that afternoon satisfying $t_1 < \text{temp} \leq t_2$ would be $\text{prob}\{t_1 \leq \text{temp} \leq t_2\} = p(t_2) - p(t_1)$. A meteorologist setting up a mathematical model of temperature variation would want to describe $p(t)$ by a continuous, in fact differentiable function. In that case, the probability of any given, single, value of the temperature would be zero, but there would be a non-zero probability of a given range or interval of temperatures.

The probabilist calls $p(t)$ the probability distribution function for the random quantity, and the derivative $dp(t)/dt$ the probability density function. Since $p(t)$ can only increase with t, it is clear that the density function will be positive, or at least non-negative. Denoting the density function by

$$\frac{d}{dt}p(t) = f(t)$$

we then have for the probability of a temperature in the range $(t_1, t_2]$ the result that

$$\text{prob}\{t_1 < \text{temp} \leq t_2\} = \int_{t_1}^{t_2} \frac{d}{dt}p(t)dt = \int_{t_1}^{t_2} f(t)dt$$

so probabilities are obtained by integrating probability density functions. You should think of the density function as a probability per unit temperature interval, in the limit as that temperature interval approaches zero. For example, the probability of a temperature between 30 and 30.1°C is actually $\int_{30}^{30.1} f(t)dt$, and is likely to be close to the value $f(30) \times 0.1$. A description of the temperature as it fluctuates randomly is then obtained either from the distribution function $p(t)$ **or** the density function $f(t)$. If $p(t)$ is known, then $f(t)$ can be determined, since

$$f(t) = \frac{d}{dt}p(t)$$

whereas if $f(t)$ is known then $p(t)$ can be determined, as

$$p(t) = \int_0^t f(s)ds$$

Here I am assuming that there is **no** chance of the temperature approaching 0°C, this being in agreement with my recent experience!

The interpretation of the definite integral as the integral of a density function can be found in many other areas of mathematics. Given a **mass distribution** along the x-axis, $\rho(x)$ will be a mass per unit volume, and the total mass between $x = x_1$ and $x = x_2$ is then $\int_{x_1}^{x_2} \rho(x)dx$. If the mass distribution is **uniform**, then ρ is a constant function, but if the distribution is denser as we move to the right then ρ will be an increasing function. The function ρ may be a density of **anything**. If ρ is a charge density, then since charges can come in either positive or negative amounts, ρ may not be of a fixed sign; or $\rho(t)$ may represent the rate at which treacle is being exported to Switzerland, as a function of time t, in which case $\int_{t_1}^{t_2} \rho(t)dt$ will describe the **total** exports of treacle to Switzerland during the time interval from t_1 to t_2.

EXERCISES ON 9.2

1. Obtain the identities, for the sine and cosine functions

$$
\left.\begin{array}{l}
\sin x = \displaystyle\int_0^x \cos t\, dt \\[2mm]
\cos x = 1 - \displaystyle\int_0^x \sin t\, dt
\end{array}\right\}
$$

Using Example 5 as a guide, and starting from the inequality $\sin x \le 1$, obtain a sequence of inequalities for the sine and cosine functions with $x > 0$, and use these to derive the Taylor series for these functions.

2. Carry out the iteration $y_{n+1}(x) = 1 - \int_0^x y_n(t)dt$ described in Example 5, starting with $y_0(x) = 1$, and verify that the iteration leads to the Taylor expansion for the function $\exp(-x)$.

   A function $f(x)$ satisfies the equation $df/dx = f^2$ for all x in the interval $-1 < x < 1$, together with the initial condition $f(0) = 1$. Show that

$$
f(x) = 1 + \int_0^x (f(t))^2 dt
$$

   Use the iterative scheme $f_{n+1}(x) = 1 + \int_0^x (f_n(t))^2 dt$, with $f_0(x) = 1$, to derive a power series expansion for $f(x)$; hence identify the function $f(x)$.

3. The mean value of a function f on an interval $[a, b]$ is defined to be $[1/(b-a)]\int_a^b f(t)dt$. Determine the mean value $m(x)$ on the interval $[0, x]$, of each of the functions f_1, f_2, defined by $f_1(t) = \cos t, f_2(t) = \cos^2 t$. In the limit as x tends to infinity, show that in the case of f_1, $m(x)$ tends to zero, whereas in the case of f_2, $m(x)$ tends to $\frac{1}{2}$.

4. A non-uniform rod of length L occupies the interval $0 \le x \le L$ of the x-axis, and has mass per unit length ρ given as a function of x by $\rho(x) = kx$, where k is a constant. Determine the total mass M of the rod. If the **moment of inertia** I of the rod about an axis through the end $x = 0$, perpendicular to the rod, is given by $I = \int_0^L x^2 \rho(x)dx$, show that $I = ML^2/2$.

5. The number of cars passing two points A,B of the northbound carriageway of a motorway, where B is 1 km to the north of A, and there are no entry or exit roads between A and B, are given, respectively, as a function of time t by $N_A(t)$ and $N_B(t)$. Write down an expression for the **increase**, over a time interval $t_1 \le t \le t_2$, in the number of cars between A and B.

   Traffic at B is brought to a halt, so that $N_B(t) = 0$, while traffic flow at A continues at a rate $N_A(t) \ge Q$. How long can this state of affairs continue before traffic along the entire stretch of motorway between A and B is brought to a halt given that this stretch of motorway can accommodate at most N_0 cars?

9.3 The improper integral

Suppose we want to determine the area between the curve $y = 1/(1 + x^2)$ and the x-axis, for positive values of x. This is the area of the shaded area in Fig 9.3. The evaluation of this area is not quite covered by the theory which I have described up to now, because the region in question is of infinite size, extending from $x = 0$ to ∞. It is not even obvious that the area of the shaded region will be finite, and if the area **is** finite we need to know how to evaluate it. Extending our previous notation in a natural way, I shall write the area as $\int_0^\infty [1/(1 + x^2)]\,dx$ (or as $\int_0^\infty [1/(1 + t^2)]\,dt$, or as $\int_0^\infty [1/(1 + w^2)]\,dw$). Such an integral is called an improper integral. An obvious way to define an improper integral like this is as the limit as N tends to infinity of the definite integral $\int_0^N [1/(1 + x^2)]\,dx$. Thus, by definition

$$\int_0^\infty \frac{1}{1 + x^2}\,dx = \lim_{N \to \infty} \int_0^N \frac{1}{1 + x^2}\,dx$$

This is to regard the area of the shaded region in Fig 9.3 as the limit as $N \to \infty$ of the area of the shaded region in Fig 9.4 between $x = 0$ and $x = N$.

An improper integral may be **convergent** or **divergent**. In this case, if the definite integral $\int_0^N [1/(1 + x^2)]\,dx$ tends to a limit as N tends to infinity, then this limit is equal to the improper integral $\int_0^\infty [1/(1 + x^2)]\,dx$, and we shall say that the improper integral is convergent. On the other hand, if $\int_0^N [1/(1 + x^2)]\,dx$ does **not** tend to a limit as N tends to infinity, we shall say that the improper integral is divergent. A fairly common convention is to write $\int_0^\infty [1/(1 + x^2)]\,dx < \infty$ if the integral converges, and $\int_0^\infty [1/(1 + x^2)]\,dx = \infty$ if the integral diverges.

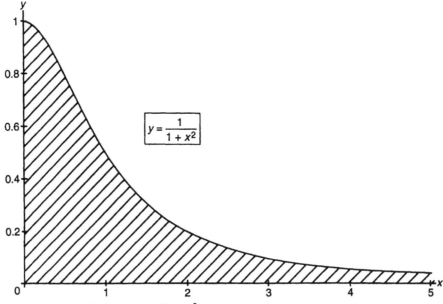

Fig 9.3 $y = 1/(1 + x^2)$, area under the curve for $x > 0$.

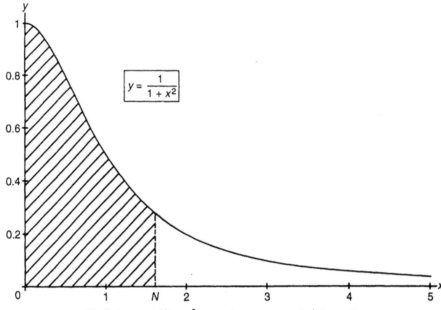

Fig 9.4 $y = 1/(1 + x^2)$, area between $x = 0$ and $x = N$.

Since an antiderivative of $1/(1 + x^2)$ is $\tan^{-1} x$, we can easily evaluate the definite integral $\int_0^N [1/(1 + x^2)]\,dx$, to give $\int_0^N [1/(1 + x^2)]\,dx = [\tan^{-1} x]_0^N = \tan^{-1} N - \tan^{-1} 0 = \tan^{-1} N$. What is the limit of $\tan^{-1} N$ as N tends to infinity? Well, if $y = \tan^{-1} N$, then $N = \tan y$ with $-\pi/2 < y < \pi/2$, and we know that $N \to \infty$ as $y \to \pi/2$. Hence $y \to \pi/2$ as $N \to \infty$. You should confirm this by considering the graphs of the functions $\tan$ and $\tan^{-1}$; so we have $\lim_{N \to \infty} \tan^{-1} N = \pi/2$.

Hence the improper integral $\int_0^\infty [1/(1 + x^2)]\,dx$ is **convergent**, and we have

$$\int_0^\infty \frac{1}{1 + x^2}\,dx = \lim_{N \to \infty} \int_0^N \frac{1}{1 + x^2}\,dx = \lim_{N \to \infty} \tan^{-1} N = \pi/2$$

There is no harm, if you are careful, in writing this calculation as $\int_0^\infty [1/(1 + x^2)]\,dx = [\tan^{-1} x]_0^\infty = (\pi/2 - 0) = \pi/2$; but $[\tan^{-1} x]_0^\infty$, here, should be understood not as $\tan^{-1} \infty - \tan^{-1} 0$, which does not make sense, but as $\tan^{-1} N - \tan^{-1} 0$ as $N \to \infty$.

An example of an improper integral which diverges is $\int_0^\infty [1/(1 + x)]\,dx$, where

$$\int_0^\infty \frac{1}{1 + x}\,dx = \lim_{N \to \infty} \int_0^N \frac{1}{1 + x}\,dx = \lim_{N \to \infty} [\ln(1 + x)]_0^N = \lim_{N \to \infty} \ln(1 + N)$$

This integral is divergent, because $\ln(1 + N)$ tends to infinity as N tends to infinity.

Improper integrals over intervals other than $(0, \infty)$ may be defined in a similar way. For example, $\int_{-\infty}^0 x e^{-x^2}\,dx$ is given by $\int_{-\infty}^0 x e^{-x^2}\,dx = \lim_{N \to \infty} \int_{-N}^0 x e^{-x^2}\,dx$. This integral is convergent, and $\int_{-N}^0 x e^{-x^2}\,dx = [-\frac{1}{2} e^{-x^2}]_{-N}^0 = (-\frac{1}{2} + \frac{1}{2} e^{-N^2})$, which converges to $-1/2$ as N tends to infinity. Hence $\int_{-\infty}^0 x e^{-x^2}\,dx = -\frac{1}{2}$.

If **both** the integrals $\int_{-\infty}^0 f(x)\,dx$ and $\int_0^\infty f(x)\,dx$ are convergent, then the improper integral $\int_{-\infty}^\infty f(x)\,dx$ over the entire real line is defined by

$$\int_{-\infty}^{\infty} f(x)dx = \int_{-\infty}^{0} f(x)dx + \int_{0}^{\infty} f(x)dx$$

$$= \lim_{N \to \infty} \int_{-N}^{0} f(x)dx + \lim_{N \to \infty} \int_{0}^{N} f(x)dx$$

$$= \lim_{N \to \infty} \left\{ \int_{-N}^{0} f(x)dx + \int_{0}^{N} f(x)dx \right\} = \lim_{N \to \infty} \int_{-N}^{N} f(x)dx$$

However, if either or both of the limits $\lim_{N \to \infty} \int_{-N}^{0} f(x)dx$ or $\lim_{N \to \infty} \int_{0}^{N} f(x)dx$ fails to exist, then we shall say that the improper integral $\int_{-\infty}^{\infty} f(x)dx$ is divergent.

A little care is needed here. Since both limits

$$\lim_{N \to \infty} \int_{-N}^{0} xe^{-x^2}dx$$

and

$$\lim_{N \to \infty} \int_{0}^{N} xe^{-x^2}dx$$

exist, the integral $\int_{-\infty}^{\infty} xe^{-x^2}dx$ is convergent, and is given by

$$\int_{-\infty}^{\infty} xe^{-x^2}dx = \lim_{N \to \infty} \int_{-N}^{N} xe^{-x^2}dx = \lim_{N \to \infty} \left[-\frac{1}{2}e^{-x^2} \right]_{-N}^{N}$$

$$= \lim_{N \to \infty} \left(-\frac{1}{2}e^{-N^2} + \frac{1}{2}e^{-N^2} \right) = 0$$

However, in the case of the integral $\int_{-\infty}^{\infty} \sin xdx$, we have

$$\int_{-N}^{0} \sin xdx = [-\cos x]_{-N}^{0} = \cos N - 1$$

and

$$\int_{0}^{N} \sin xdx = [-\cos x]_{0}^{N} = 1 - \cos N$$

Since $\cos N$ is an oscillating function which does not tend to a limit as N tends to infinity, **neither** of the limits $\lim_{N \to \infty} \int_{-N}^{0} \sin xdx$ nor $\int_{0}^{N} \sin xdx$ tends to a limit, and we have to say that $\int_{-\infty}^{\infty} \sin xdx$ is a divergent integral. This is so despite the fact that $\lim_{N \to \infty} \int_{-N}^{N} \sin xdx = 0$ which **does** have a limit as N tends to infinity. The point is that, for improper integrals as well as standard definite integrals, we must insist on basic properties of the integral (here $\int_{-\infty}^{\infty} f(x)dx = \int_{-\infty}^{0} f(x)dx + \int_{0}^{\infty} f(x)dx$) and we cannot cancel one divergent integral with another.

The idea of an improper integral extends in many cases to integrals over **finite** intervals, particularly in the case of integrals $\int_{a}^{b} f(x)dx$ where the function f has an antiderivative at every point x of the **open** interval $a < x < b$, but there is no antiderivative at one or other (or both) of the endpoints $x = a, x = b$. As a simple example, note that the function $1/\sqrt{x}$ has an antiderivative $2\sqrt{x}$ for every $x > 0$, but **not** at $x = 0$. Hence we may define an improper integral $\int_{0}^{1}(1/\sqrt{x})dx$ by $\int_{0}^{1}(1/\sqrt{x})dx = \lim_{N \to 0+} \int_{N}^{1}(1/\sqrt{x})dx$, where the limit $N \to 0$ must be taken through positive values. This gives

$$\int_0^1 \frac{1}{\sqrt{x}} dx = \lim_{N \to 0+} [2\sqrt{x}]_N^1 = \lim_{N \to 0+} 2(1 - \sqrt{N}) = 2$$

Here are some further examples of improper integrals.

● Example 6

Consider the integral $\int_0^\infty x^2 e^{-x} dx$. This improper integral is given by

$$\lim_{N \to \infty} \int_0^N x^2 e^{-x} dx$$

which on integration by parts gives

$$\lim_{N \to \infty} [-(x^2 + 2x + 2)e^{-x})]_0^N = \lim_{N \to \infty} \{2 - (N^2 + 2N + 2)e^{-N}\}$$

We have already seen that $\lim_{N \to \infty} N^k e^{-N} = 0$ for any $k > 0$. Hence the integral is convergent, and we find $\int_0^\infty x^2 e^{-x} dx = 2$.

Noting that $\lim_{x \to \infty} (x^2 + 2x + 2)e^{-x} = 0$, this integral may also be set out as

$$\int_0^\infty x^2 e^{-x} dx = [-(x^2 + 2x + 2)e^{-x}]_0^\infty = 0 + 2 = 2$$

● Example 7

We consider the two integrals

$$I_1 = \int_1^\infty \frac{t}{1 + t^4} dt, \quad I_2 = \int_1^\infty \frac{t^3}{1 + t^4} dt$$

Then $I_1 = \lim_{N \to \infty} \int_1^N [t/(1 + t^4)] dt$; making the change of integration variable $t = \sqrt{s}$, $dt = (dt/ds) ds = (1/2\sqrt{s}) ds$, we have

$$I_1 = \lim_{N \to \infty} \int_{t=1}^{t=N} \frac{t}{1 + t^4} dt = \lim_{N \to \infty} \int_1^{N^2} \frac{\sqrt{s}}{1 + s^2} \cdot \frac{1}{2\sqrt{s}} ds$$

$$= \lim_{N \to \infty} \int_1^{N^2} \frac{1}{2(1 + s^2)} ds = \lim_{N \to \infty} \left[\frac{1}{2} \tan^{-1} s \right]_1^{N^2}$$

$$= \lim_{N \to \infty} \left\{ \frac{1}{2} (\tan^{-1} N^2 - \tan^{-1} 1) \right\}$$

$$= \frac{1}{2} \left(\frac{\pi}{2} - \frac{\pi}{4} \right) = \frac{\pi}{8}$$

Hence $\int_1^\infty [t/(1 + t^4)] dt = \pi/8$. This integral can also be evaluated as

$$\int_{t=1}^{t=\infty} \frac{t}{1 + t^4} dt = \int_{s=1}^{s=\infty} \frac{1}{2(1 + s^2)} ds = \left[\frac{1}{2} \tan^{-1} s \right]_1^\infty = \frac{\pi}{8}$$

Here we have made use of the fact that $t \to \infty \Rightarrow s \to \infty$, and that $\lim_{s \to \infty} \tan^{-1} s = \pi/2$.

In the case of the second integral I_2, observe that the numerator t^3 of the integrand is also the derivative of the denominator $1 + t^4$, apart from a multiplicative constant. This leads to an antiderivative $\frac{1}{4}\ln(1 + t^4)$ (or alternatively, make a change of variable $1 + t^4 = s$), and we have $\int_1^\infty [t^3/(1 + t^4)]\,dt = \lim_{N\to\infty}[\frac{1}{4}\ln(1 + t^4)]_1^N$. Since $\ln(x)$ tends to infinity for large x, the integral diverges, and we have $\int_1^\infty [t^3/(1 + t^4)]\,dt = \infty$.

Had we not been able to find an antiderivative, it would still be possible to **compare** this integral with another integral which we **could** evaluate. For example, $t^3/(1 + t^4) \geq 1/2t$ for $t \geq 1$, and since $\int_1^N (1/t)\,dt = \ln(N)$ tends to infinity, it follows that $\lim_{N\to\infty}\int_1^N [t^3/(1 + t^4)]\,dt \geq \lim_{N\to\infty}\int_1^N (1/2t)\,dt$ diverges also.

❋ Example 8

Two well-known examples of improper integrals are $I_1 = \int_{-\infty}^\infty e^{-x^2}dx$ and $I_2 = \int_0^\infty (\sin x/x)\,dx$. Although we can verify that both of these integrals are convergent, their evaluation is beyond the scope of methods described in this book. (I_1 is evaluated by expressing I_1^2 as a double integral in two-dimensional space, and making a change of variables to plane polar coordinates; I_2 may be evaluated by techniques involving complex contour integrals).

In considering an integral such as $\int_0^\infty e^{-x^2}dx$, which may be expressed as a limit $\int_0^\infty e^{-x^2}dx = \lim_{N\to\infty}\int_0^N e^{-x^2}dx$, notice that $\int_0^N e^{-x^2}dx$ is an increasing function of N; in fact $d/dN \int_0^N e^{-x^2}dx = e^{-N^2} > 0$. Hence $\int_0^N e^{-x^2}dx$ will either increase without limit, in which case $\int_0^\infty e^{-x^2}dx = \infty$, or will remain bounded as N increases, in which case the integral will converge to a limit as $N \to \infty$. A simple comparison shows that the integral is bounded, since $e^{-x^2} \leq 1$ for $0 \leq x \leq 1$ and $e^{-x^2} \leq e^{-x}$ for $x \geq 1$, so that for $N > 1$ we have $\int_0^N e^{-x^2}dx = \int_0^1 e^{-x^2}dx + \int_1^N e^{-x^2}dx \leq \int_0^1 1\,dx + \int_1^N e^{-x}dx \leq 2 - e^{-N} < 2$. So $\int_0^\infty e^{-x^2}dx$ is convergent. Similarly $\int_{-\infty}^0 e^{-x^2}dx$ is also convergent, and in fact it is easy to show that $\int_{-\infty}^0 e^{-x^2}dx = \int_0^\infty e^{-x^2}dx$. It follows that $I_1 = \int_{-\infty}^\infty e^{-x^2}dx$ is a convergent integral. It may be shown that $I_1 = \sqrt{\pi}$.

In the second example, $I_2 = \int_0^\infty (\sin x/x)\,dx$, the integrand is undefined at $x = 0$. Nevertheless, the inequality $\sin x \leq x$ for $x > 0$ (see problem 1 of Exercises on 9.2) shows that, for example, $0 \leq \int_0^1 (\sin x/x)\,dx = \lim_{N\to 0+}\int_N^1 (\sin x/x)\,dx \leq \lim_{N\to 0+}\int_N^1 (1)\,dx = 1$, so the improper integral converges over any **finite** positive interval. It remains to consider the integral $\int_0^N (\sin x/x)\,dx$ in the limit $N \to \infty$. Writing this integral, for $N > 1$, as $\int_0^1 (\sin x/x)\,dx + \int_1^N (\sin x/x)\,dx$, the second integral may be integrated by parts to give

$$\int_1^N \frac{\sin x}{x}\,dx = \left[-\frac{\cos x}{x}\right]_1^N - \int_1^N \frac{\cos x}{x^2}\,dx$$

which converges in the limit to $\cos 1 - \int_1^\infty (\cos x/x^2)\,dx$. Here the integrand $\cos x/x^2$ lies between $-1/x^2$ and $+1/x^2$, and since $\int_1^\infty (1/x^2)\,dx = [-1/x]_1^\infty = 1$ is a convergent integral, it may be verified that $\int_1^\infty (\sin x/x)\,dx$ is also convergent. Hence $I_2 = \int_0^\infty (\sin x/x)\,dx$ is convergent, and it may be shown that I_2 has the value π.

EXERCISES ON 9.3

1. Which of the three integrals

$$\int_0^\infty \frac{x}{1+x}\,dx, \quad \int_0^\infty \frac{x}{(1+x)^2}\,dx, \quad \int_0^\infty \frac{x}{(1+x)^3}\,dx$$

converge? Evaluate any of the integrals which is convergent.

2. Use integration by parts to show that an antiderivative of $(\ln(x))^2$ for $x > 0$ is $x[(\ln(x))^2 - 2\ln(x) + 2]$. Hence verify that the improper integral $\int_0^1 (\ln(x))^2 dx$ has the value 2.

3. Use the derivative of an inverse trigonometric function to show that $\int_0^1 (1/\sqrt{1-x^2})\,dx = \pi/2$. Verify the identity

$$\sqrt{1-x^2} = \frac{1}{2}\left\{ \frac{d}{dx}(x\sqrt{1-x^2}) + \frac{1}{\sqrt{1-x^2}} \right\} \quad (-1 < x < 1)$$

and hence evaluate the integral $\int_0^1 \sqrt{1-x^2}dx$.

Summary

The definite integral $\int_a^b f(x)dx$ is the area between the curve $y = f(x)$ and the x-axis over the interval $a \leq x \leq b$, with proper account taken of the sign according as the curve is above or below the x-axis. If f is continuous on the closed interval $a \leq x \leq b$, then the definite integral is well defined, and is given by $\int_a^b f(x)dx = [F(x)]_a^b = F(b) - F(a)$, where F is an antiderivative of f on the closed interval. The fundamental theorem of calculus then says that $(d/dx)\int_a^x f(t)dt = f(x)$, for $a \leq x \leq b$, so that $F(x) = \int_a^x f(t)dt$ is an antiderivative. The definite integral depends linearly on the integrand f, in the sense that $\int_a^b (Af(x) + Bg(x))dx = A\int_a^b f(x)dx + B\int_a^b g(x)dx$, for any constants A, B. The definite integral depends monotonically on the function f, in the sense that if $f \geq g$ then $\int_a^b f(x)dx \geq \int_a^b g(x)dx$. Besides its interpretation as area, the definite integral has a number of other possible meanings in which the integrand is a density and the integral itself may be interpreted as total mass, charge, probability, and so on.

Integrals such as $\int_0^\infty f(x)dx$, or $\int_a^b f(x)dx$ where a or b is a point of non-continuity of f, may often be evaluated as improper integrals; for example $\int_0^\infty f(x)dx = \lim_{N\to\infty} \int_0^N f(x)dx$ or $\int_a^b f(x)dx = \lim_{N\to b^-} \int_a^N f(x)dx$, where the improper integral is said to converge whenever the appropriate limit exists, and diverge otherwise.

FURTHER EXERCISES

1. Use integration by parts, with

$$\frac{x^2}{(1+x^2)^2} = -\frac{x}{2}\frac{d}{dx}\frac{1}{1+x^2}$$

to show that

$$\int_0^1 \frac{x^2}{(1+x^2)^2}\,dx = \frac{\pi-2}{8}$$

Hence use the identity

$$\frac{1}{(1+x^2)^2} = \frac{1}{1+x^2} - \frac{x^2}{(1+x^2)^2}$$

to evaluate the integral $\int_0^1 [1/(1+x^2)^2]\,dx$.

   Determine also the improper integral $\int_0^\infty [1/(1+x^2)^2]\,dx$, and verify your result by means of the change of variable $x = \tan\theta$.

2. Find an antiderivative of $e^{-3x}\sin x$ of the form $e^{-3x}(a\cos x + b\sin x)$, and hence evaluate the integral $\int_0^\infty e^{-3x}\sin x\,dx$.

3. Use partial fractions to evaluate the integral $\int_1^\infty [1/t(t+1)]\,dt$, and verify your results by means of the change of variable $t = 1/s$.

4. Use partial fractions to determine each of the integrals

$$\int_0^1 \frac{x}{(x+1)(x+2)}\,dx, \quad \int_0^1 \frac{x}{(x+1)(x^2+1)}\,dx \text{ and } \int_0^\infty \frac{1}{(x+1)(x^2+1)}\,dx$$

5. Use integration by parts to evaluate each of the integrals

$$\int_0^\pi x^2\sin 2x\,dx, \quad \int_0^\infty x^2 e^{-2x}\,dx \text{ and } \int_0^1 x^2\ln(x)\,dx$$

6. Show that the integral $\int_0^\infty [1/\sqrt{x}(1+x)]\,dx$ is convergent, and make a change of integration variable to evaluate the integral. Show that the integral $\int_0^\infty [1/\sqrt{x}(1+\sqrt{x})]\,dx$ is divergent.

7. Verify the identity

$$\frac{d}{dx}\{x\sqrt{x^2-1}\} = 2\sqrt{x^2-1} + \frac{1}{\sqrt{x^2-1}} \quad (|x|>1)$$

and hence use an inverse hyperbolic function to evaluate the integral $\int_1^{\sqrt{2}} \sqrt{x^2-1}\,dx$.

8. Given an odd function f (so that $f(-t) = -f(t)$), make the change of variable $t = -s$ to show that $\int_{-\pi}^{\pi} f(t)\,dt = -\int_{-\pi}^{\pi} f(s)\,ds$. Hence show that $\int_{-\pi}^{\pi} f(t)\,dt = 0$ for any odd function f, and give a geometrical explanation for this result.

9. Assuming the result $(d/dx)\int_c^x f(t)\,dt = f(x)$ for any constant c, show that

$$\frac{d}{dx}\int_x^c f(t)\,dt = -f(x)$$

and obtain the identity

$$\frac{d}{dx} \int_{-x}^{x} f(t)dt = f(x) + f(-x)$$

10. For a particle moving along the x-axis, the acceleration d^2x/dt^2 is given as a function of time t by $d^2x/dt^2 = f(t)$. At time $t = 0$, the particle is at rest at the origin, i.e. $x = 0$ and $dx/dt = 0$ at $t = 0$. Show that the displacement of the particle is given as a function of time by $x(t) = t \int_0^t f(s)ds - \int_0^t sf(s)ds$. Determine $x(t)$ in the case $f(t) = a \sin \omega t$, where a and ω are constants.

10 • Differential Equations

The subject of this chapter is not altogether new, because we have actually already met several **examples** of differential equations. The equation $df/dx = f$, which was solved for f in Section 7.1, is an example of a **first order differential equation**. First order because the equation involves a first order derivative df/dx, rather than higher order derivatives of f. The exponential function is said to be a **solution** of this differential equation, or to **satisfy** the differential equation, because if $f(x) = \exp(x)$ then $df/dx = f$ amounts to $d(\exp(x))/dx = \exp(x)$, which holds for all values of x. The exponential function is not the **only** solution of this differential equation, since any constant multiple of the exponential function is also a solution:

$$\frac{d}{dx}\{A \exp(x)\} = A \exp(x)$$

In Section 7.1 we showed that **every** solution of $df/dx = f$ is a constant multiple of the exponential function; there are no other solutions. The **set** of functions satisfying $df/dx = f$ is called the **solution set** of the differential equation, so the solution set of $df/dx = f$ consists of all functions which are a constant multiple of $\exp(x)$. In set notation, the solution set is $\{A \exp(x);\ A \in \mathbb{R}\}$, and this is a set of functions. We also say that the **general solution** of $df/dx = f$ is $f(x) = A \exp(x)$.

Another very simple example of a differential equation, is the equation $df/dx = \sin x$. This is the equation satisfied by any antiderivative of the sine function. Certainly $f(x) = -\cos x$ is a solution of $df/dx = \sin x$, since $d(-\cos x)/dx = \sin x$; but $f(x) = c - \cos x$ is also a solution, for any constant (real number) c. Again, we know that every solution is of the form $c - \cos x$ for some constant c, so we can say in this case that the solution set consists of $\{c - \cos x; c \in \mathbb{R}\}$, or that $f(x) = c - \cos x$ is the general solution of $df/dx = \sin x$.

Here, too, $df/dx = \sin x$ is a first order differential equation. What would a second order differential equation look like? One example of a second order equation is $d^2x/dt^2 = \sin t$, and you have already seen examples of this type in the exercises at the end of Chapter 9. Here the equation is to be solved for x as a function of t. This is a second order equation because a second order derivative of x is involved, but no higher order derivatives. One solution in this case is $x(t) = -\sin t$. Certainly this is not the only solution, and you may care to convince yourself that the general solution is $x(t) = A + Bt - \sin t$.

Here are some other examples of differential equations: $df/dx = f^2$ (first order); $df/dx = x^2 + f^2$ (first order); $d^3f/dx^3 + x^2d^2f/dx^2 = f^2$ (third order); $d^2f/dx^2 = x^2\cos(f)$ (second order); $d^4u/dt^2 + u = \cos t$ (fourth order, to be solved for u as a function of t). The time has come to see what all these examples have in common and to establish general methods of dealing with differential equations. You will find that the theory is one of the deepest and most fruitful applications of some of the ideas which have been occupying us in previous chapters.

10.1 First order differential equations

It will help to start with at least a working definition of what we mean by the term 'differential equation'. At the same time, I shall try to make rather more precise some of the terminology. In the following definition, I want to emphasize that, for us, the solution of a differential equation will be **on an interval**. That is, it does not make sense to refer to a solution of a differential equation **at a point**, or on a union of several intervals, or on more complicated subsets of $\mathbb{R}$ — just on an interval will do nicely.

● *Definition 1*

A first order differential equation for a function $f(x)$ is an equation of the form $df/dx = Q(x,f)$ (or an equation reducible to this form), where Q is a **given** function of x and f. Examples are: $df/dx = x^2 + f^2$, $df/dx = \cos(xf)$, $df/dx = \sin x$ and $df/dx = f^2$.

A function f is said to be a **solution** of the differential equation on an interval $a < x < b$ if

$$\frac{d}{dx} f(x) = Q(x,f(x))$$

for all x in the interval $a < x < b$. For example, a solution $f(x)$ of $df/dx = x^2 + f^2$ on the interval $0 < x < 1$ must satisfy $df(x)/dx = x^2 + (f(x))^2$ for all x on this interval; solutions on other intervals, including closed intervals, may also be considered. The function Q is assumed to have continuous dependence on x and f as x varies over the interval in question.

The set of all functions satisfying the differential equation (i.e. all solutions of the equation) on a given interval, is called the **solution set** of the equation on that interval, and any general expression for a function in that solution set may be described as the **general solution** of the equation.

I have left what is meant by the 'continuous dependence' of the function Q on x and f a little vague, since it goes a little beyond the scope of this book to introduce continuity for functions of two or more variables (functions from $\mathbb{R}^n$ to $\mathbb{R}$ with $n \geq 2$). The important point is that there must be some functional relationship which gives the value of df/dx once we are told the values of x and f, and this functional relationship is usually assumed to be smooth in some sense, in order to avoid certain pathologies which might otherwise occur. In most cases, the functional dependence of df/dx on x and f involves some combination of the standard functions which we have seen already.

What does it **mean** to write down a first order differential equation (or DE for short)? Evidently, the derivative df/dx must be some given function, Q, of x and f; $df/dx = Q(x,f)$. Any function f which satisfies this equation, for all x in an interval $a < x < b$, is a **solution** of the DE. Given any solution, let $y = f(x)$ be the equation of the graph of the solution, in the interval $a < x < b$. A graph of a solution of the DE is called a **solution curve** of the DE. Suppose, then, that the solution curve passes through the point having Cartesian coordinates (x_0, y_0), with $a < x_0 < b$. In that case, when $x = x_0$ we have $y = f(x_0) = y_0$. Since, by assumption, $df/dx = Q(x,f)$ for **all** x in the interval (a,b), it follows in particular that at

$x = x_0$, $y = y_0$ we have $df/dx = Q(x_0, f(x_0)) = Q(x_0, y_0)$. This gives a geometrical interpretation of the DE. At any point (x_0, y_0) of a solution curve, the DE tells us what the gradient of the solution curve must be. Namely if a solution curve passes through (x_0, y_0), then it does so with gradient $Q(x_0, y_0)$. To summarize:

A solution of the DE is a function f, which satisfies

$$\frac{df}{dx} = Q(x, f) \text{ for all } x \in (a, b)$$

The graph of this solution is a solution curve $y = f(x)$, which satisfies

$$\frac{dy}{dx} = Q(x, y) \text{ at any point } (x, y) \text{ of the graph}$$

so a solution curve passing through any point (x_0, y_0) must have gradient $Q(x_0, y_0)$ at that point.

One more important property of solution curves of first order DEs, which holds whenever the function $Q(x, y)$ is defined at (x_0, y_0) and has reasonably smooth dependence on x and y (I shall be more precise later), is the following.

There is one and only one solution curve passing through any given point (x_0, y_0). This solution curve is the graph $y = f(x)$ of a solution f of the DE subject to the condition $f(x_0) = y_0$, so $y = y_0$ when $x = x_0$. The condition $f(x_0) = y_0$, or $y = y_0$ when $x = x_0$, is described as an **initial condition**. The fact that there is just one solution curve through a given point corresponds analytically to the statement that there is just one solution of the first order DE satisfying a given initial condition.

A few examples will help to make some of these ideas clearer.

Example I

We consider the first order DE $df/dx = x + f$. This DE is of the form $df/dx = Q(x, f)$, where the function Q is given by $Q(x, f) = (x + f)$. In terms of the graph or solution curve $y = f(x)$, we might prefer to write $dy/dx = x + y$.

How can we solve this DE? In fact, we have already solved a very similar equation in Section 7.1, where we obtained the general solution of $df/dx = f$, namely $f(x) = Ae^x$. Our method in that case was to begin by considering the derivative $d\{e^{-x}f\}/dx$, and the same idea will work in this example too.

Suppose, then, that a function f satisfies the first order DE $df/dx = x + f$. By the product rule for differentiation, we have

$$\frac{d}{dx}\{e^{-x}f\} = e^{-x}\frac{df}{dx} + f\frac{d}{dx}(e^{-x}) = e^{-x}\left(\frac{df}{dx} - f\right)$$

From the DE, $df/dx - f = x$. Hence

$$\frac{d}{dx}\{e^{-x}f\} = xe^{-x}$$

Knowing that the derivative of $e^{-x}f$ is xe^{-x}, it follows that $e^{-x}f$ must be an antiderivative of xe^{-x}. Thus $e^{-x}f = \int xe^{-x}dx + c$, where c is a constant. Using integration by parts to evaluate the integral, this leads to $e^{-x}f = -(x+1)e^{-x} + c$;

so $f(x) = ce^x - (1 + x)$. This is the general solution to the DE. That is, not only is $f(x) = ce^x - (1 + x)$ a solution of the DE, but every solution of the DE is of this form, for some value of the constant c. The solution set of the DE is the set of functions $\{ce^x - (1 + x)\}$, as c is varied. For each value of c, $y = ce^x - (1 + x)$ is the equation of a solution curve, or graph of the solution corresponding to this value of c.

It is instructive to sketch these various solution curves, for different values of c, on a single diagram. The simplest case is $c = 0$, for which we have the straight line graph $y = -(1 + x)$. For $c = 1$, we have the solution curve $y = e^x - (1 + x)$, which passes through $(0,0)$ with zero gradient. For $c = 2$, the solution curve $y = 2e^x - (1 + x)$ passes through the point $(0,1)$, with gradient 1. Every solution curve asymptotically approaches the straight line $y = -(1 + x)$ in the limit as x tends to $-\infty$, and diverges exponentially in the limit $x \to \infty$, except in the case $c = 0$.

Some of these solution curves are sketched in Fig 10.1. Note the role of the DE in specifying the gradient at any point of a solution curve. For example, we know without having to solve the DE that any solution curve passing through the point $x = 0$, $y = 1$ must have a gradient, at that point, given by $dy/dx = x + y = 1$. As we have seen, the solution curve through $(0,1)$ is $y = 2e^x - (1 + x)$, and it is easy to verify that we have the correct gradient at this point.

I have said that we can expect one and only one solution curve through any given point. What is the solution curve through the point (x_0, y_0)? The answer to this question is equivalent to determining the solution $f(x)$ of the DE, subject to a given initial condition $f(x_0) = y_0$.

Since **every** solution of the DE is of the form $ce^x - (1 + x)$ for some constant c, we know that the solution curve through (x_0, y_0) must be of the form

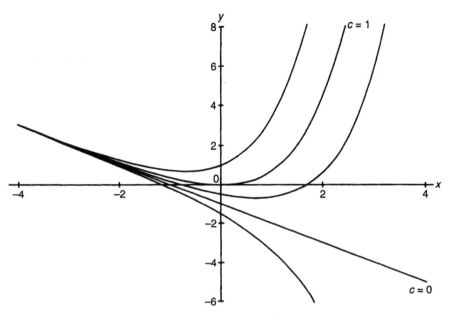

Fig 10.1 Solution curves of $df/dx = x + f$.

$y = ce^x - (1 + x)$, where c is to be determined. Since $y = y_0$ at $x = x_0$, it also follows that $y_0 = ce^{x_0} - (1 + x_0)$. This determines the constant c, which is given in terms of x_0 and y_0 by $c = e^{-x_0}(x_0 + y_0 + 1)$. Hence the solution curve through (x_0, y_0) has the equation $y = e^{(x-x_0)}(x_0 + y_0 + 1) - (1 + x)$.

It is straightforward to verify that this equation satisfies not only $y = y_0$ at $x = x_0$, but also $dy/dx = x_0 + y_0$ at (x_0, y_0), as required. Moreover, the solution $f(x) = e^{(x-x_0)}(x_0 + y_0 + 1) - (1 + x)$ satisfies both the DE and the initial condition $f(x_0) = y_0$.

Example 2

A second, rather different, example of a first order DE is the equation $xdf/dx = 1 + f^2$. Let us solve this DE, subject to the initial condition $f(1) = 0$. Notice first of all that this DE is not yet in the standard form $df/dx = Q(x,f)$ for first order equations. To convert the equation to this form, it is necessary only to divide throughout by a factor x, to give $df/dx = (1/x)(1 + f^2)$. This is now the standard first order DE, with $Q(x,f) = (1/x)(1 + f^2)$. However, a singularity has been introduced at $x = 0$. Since $Q(x,f)$ is then undefined at $x = 0$, the DE cannot be written down in any interval of x which includes the origin. This means that we can look for solutions in an interval to the right of the origin, $a < x < b$ with a and b positive, or solutions in an interval to the left of the origin, with a and b negative, but we cannot consider solutions in any interval $a < x < b$ with a negative and b positive. Since the given initial condition refers to the point $x = 1$, it makes sense to take an interval to the right of the origin. This interval must of course contain the point $x = 1$.

The DE in this example, $df/dx = (1/x)(1 + f^2)$, for which the function $Q(x,f)$ is a product of a function of x with a function of f, belongs to a class of first order DEs referred to as **separable**. A first order DE will be called separable if it may be written in the form $df/dx = Q_1(x).Q_2(f)$, for given functions Q_1 and Q_2, where Q_1 depends only on x and Q_2 depends only on f. The following technique, applied to the current example, is often used in such cases. Examine the argument carefully.

> Suppose $df/dx = (1/x)(1 + f^2)$. Then a little simple algebra gives $df/(1 + f^2) = (1/x)dx$, where we have manipulated the equation so as to have only f on the left-hand side, and only x on the right-hand side. Insert an integral sign in front of both sides of the equation, so that $\int df/(1 + f^2) = \int (1/x)dx$. The right-hand side is easy: $\int (1/x)dx = \ln(x) + c$. The left-hand side is also a standard integral: $\int dy/(1 + y^2) = \tan^{-1} y + c$; so $\tan^{-1} f + \text{const} = \ln(x) + \text{const}$, implying that $\tan^{-1} f = \ln(x) + c$, for some constant c. Hence $f(x) = \tan(\ln(x) + c)$ is the general solution of the DE. The solution with $f(1) = 0$ gives $c = 0$, so that $f(x) = \tan(\ln(x))$.

No doubt you were as shocked by this argument as I was! The method outlined above, though in common use to solve separable DEs, simply does not stand up to close scrutiny.

Although df/dx has a precise meaning, as the derivative of the function f, we have no meaning for 'df' and 'dx' separately. Even could we do so, we should hardly expect to manipulate these expressions as algebraic quantities. (In fact, it **is** possible to give a meaning to such expressions as 'df' and 'dx' through the theory of differential forms, but this is a highly developed and advanced branch of mathematics which is hardly relevant within the present context.) It is also unclear what can be meant by $\int df/(1+f^2)$ where f is a function of x rather than an integration variable.

Although this method for solution of separable DEs cannot be considered mathematically correct, a valid method of solution **can** be constructed along somewhat similar lines. I hope that you will agree that the following argument, applied to the same DE, is both mathematically convincing **and** leads to the required solution of the problem.

Suppose $df/dx = (1/x)(1+f^2)$. If we write $y = f(x)$, then the DE looks like $dy/dx = (1/x)(1+y^2)$, or $[1/(1+y^2)]\,dy/dx = 1/x$. (This equation separates y from x as far as possible algebraically, without 'splitting' dy/dx.) The factor $1/(1+y^2)$ on the left-hand side has an antiderivative

$$\int \frac{dy}{1+y^2} = \tan^{-1} y$$

This is an antiderivative of $1/(1+y^2)$ in the sense that

$$\frac{d}{dy} \tan^{-1} y = \frac{1}{1+y^2}$$

The chain rule, with $y = f(x)$, now gives

$$\frac{d}{dx} \tan^{-1} f(x) = \frac{d}{dx} \tan^{-1} y = \left(\frac{d}{dy} \tan^{-1} y\right) \cdot \frac{dy}{dx} = \frac{1}{(1+y^2)} \frac{dy}{dx}$$

which is just $1/x$, from the DE. Hence

$$\frac{d}{dx} \tan^{-1} f(x) = \frac{1}{x}$$

from which it follows on integration that $\tan^{-1} f(x) = \ln(x) + c$, where c is a constant. So $f(x) = \tan(\ln(x) + c)$, and again from the initial condition we have $c = 0$, leading to the required solution

$$f(x) = \tan(\ln(x))$$

Notice that the solution $f(x) = \tan(\ln(x))$ is defined on the interval $\exp(-\pi/2) < x < \exp(\pi/2)$. For each value of x in this interval, the DE is satisfied. We can once again use the chain rule to verify this, since with $f(x) = \tan g(x)$ and $g(x) = \ln(x)$ we have

$$\frac{d}{dx} f(x) = (1 + \tan^2 g(x)) \cdot \frac{dg(x)}{dx} = \frac{1}{x}(1 + \tan^2 g(x)) = \frac{1}{x}(1 + (f(x))^2)$$

as required. Moreover, the initial condition is satisfied, since $f(1) = \tan(\ln(1)) = \tan 0 = 0$.

As x is increased from its initial value of one, we reach a singularity of $f(x)$ at $x = \exp(\pi/2)$, because as x approaches this value $\ln(x)$ approaches $\pi/2$ which is a singularity of the tan function. On the other hand, if we decrease x from its initial value of one, $\ln(x)$ approaches $-\pi/2$ at $x = \exp(-\pi/2)$, and again the solution diverges. Hence $(\exp(-\pi/2), \exp(\pi/2))$ is the **largest interval** on which the solution to the DE exists, subject to the initial condition $f(1) = 0$.

This kind of behaviour is typical of the solution of first order DEs, subject to an initial condition. Provided that Q has sufficiently regular behaviour as a function of x and f, the DE $df/dx = Q(x,f)$ will have a unique solution subject to a given initial condition. This solution will not necessarily be defined for all $x \in \mathbb{R}$, but there will be a maximal interval $a < x < b$, containing the point x_0 at which the initial condition is imposed, such that the DE will be satisfied for all x in this interval. The precise location of the endpoints a, b of this interval of definition of the solution will depend in general on the initial condition and on a detailed analysis of the equation. However, any singularity (point of discontinuity, point not in the domain) of the function $Q(x,f)$, say at $x = x_s$, may give rise to a corresponding singularity of the solution f itself — such singular points are often encountered as endpoints of the interval of definition. The following example shows that, even if Q varies smoothly with x and f, the interval of definition of a solution need not be the entire real line.

Example 3

We consider the DE $df/dx = f^2$, subject to initial condition $f(1) = 1$, and the DE $df/dx = (x+f)^2$, subject to $f(0) = 0$.

In the first case, following the method of Example 2, we write $y = f(x)$ and $dy/dx = y^2$. Then $(1/y^2)\, dy/dx = 1$, provided $y \neq 0$, where the factor $1/y^2$ on the left-hand side has antiderivative $-1/y$. We therefore apply the chain rule to $d(-1/y)/dx$, with $y = f(x)$. This gives

$$\frac{d}{dx}\left(-\frac{1}{f(x)}\right) = \frac{1}{(f(x))^2}\frac{df(x)}{dx} = 1$$

from the DE. Since

$$\frac{d}{dx}(1/f(x)) = -1$$

it follows on integration that $1/f(x) = -x + c$, where c is a constant. Hence

$$f(x) = \frac{1}{c - x}$$

is the general solution to the DE. [There is actually one other solution not included in this 'general solution': this is the zero function $f(x) = 0$. The zero function is trivially verified to satisfy the DE, but was omitted since we made the assumption $f(x) \neq 0$. One may show that a solution which is zero at **any** point, say at $x = x_0$, is zero at **every** point and is therefore equal to the zero function. The solution set of $df/dx = f^2$ consists of all functions $1/(c - x)$ ($c \in \mathbb{R}$) together with the zero function.]

Applying the initial condition $f(1) = 1$, we have $c = 2$, and the appropriate solution is $f(x) = 1/(2 - x)$. This solution is defined on the interval $-\infty < x < 2$.

Our second DE in this example, $df/dx = (x + f)^2$, may be cast into a more tractable form by means of the substitution $x + f(x) = g(x)$. Then

$$\frac{dg}{dx} = 1 + \frac{df}{dx} = 1 + g^2$$

Writing this DE for g in the form $[1/(1 + g^2)]\, dg/dx = 1$, we now have

$$\frac{d}{dx} \tan^{-1} g = 1$$

So $\tan^{-1} g = x + c$ for some constant c, implying $g(x) = \tan(x + c)$. The initial condition $f(0) = 0$ also gives $g(0) = 0$. Hence $c = 0$, and $g(x) = \tan x$. The solution of the DE for f is then $f(x) = \tan x - x$, and the interval of definition of this solution is $(-\pi/2, \pi/2)$.

Our final example in this section shows that things **can** go wrong for certain functions Q.

● *Example 4*

We consider the first order DE $df/dx = 3f^{2/3}$. Assuming $f(x) \neq 0$, this equation may be written as $(1/3f^{2/3})\, df/dx = 1$, which by our previous method becomes

$$\frac{d}{dx}(f^{1/3}) = 1$$

Hence $f^{1/3} = x + c$ for some constant c, which gives $f(x) = (x + c)^3$. All well and good, **provided** $f(x)$ is never zero. However, whatever the value of the constant c, we are sure to have some value of x at which $f(x) = 0$, namely $x = -c$, so the starting assumption $f(x) \neq 0$ will not then be satisfied over the entire domain.

We can see the problem in its starkest form if we ask the question: what is the solution of the DE, subject to the initial condition $f(0) = 0$? Taking $c = 0$ in the 'general solution', we arrive at the solution $f(x) = x^3$ subject to $f(0) = 0$, and it is easily verified that this function **is** a solution of the DE, and **does** satisfy the prescribed initial condition. But there is another solution, $f(x) = 0$, the zero function, which clearly **also** satisfies the DE and initial condition. In fact, you may like to verify that there are infinitely many solutions of this DE, all satisfying the initial condition $f(0) = 0$! One such solution is

$$\left. \begin{array}{ll} f(x) = 0, & -\infty < x < 1 \\ \quad\ = (x - 1)^3, & 1 \leq x < \infty \end{array} \right\}$$

The equation in this example, $df/dx = 3f^{2/3}$, fits very well into the general theory of first order DEs as regards **existence** of solutions, but not so well as regards **uniqueness**. With respect to existence, there is nothing to complain of; given any initial condition $f(x_0) = y_0$, there will always exist a solution $f(x)$ of the DE subject to this initial condition. The problem is that this solution is not unique — there is **more than one** solution subject to a given initial condition.

This is a pity. One of the nicest features of first order DEs, and one which we shall see later extended to equations of higher order, is existence **and** uniqueness for a wide class of equations. We need to know that each initial condition leads to one and only one solution, and this is very important to many of the applications. Geometrically, there has to be just one solution curve passing through any point (x_0, y_0).

Fortunately, a proof of existence and uniqueness can be found, though clearly the proof will not work for the last example, at least in the case of uniqueness. In fact, there are many such proofs, starting from a variety of different assumptions and using a variety of methods. The point is that mathematical results, if they are to be stated with precision and proved with rigour, rarely come free of charge. It is possible to prove that $df/dx = Q(x, f)$ has a unique solution subject to $f(x_0) = y_0$, but a price must be exacted in the form of assumptions to be made regarding the function Q. That is, if Q satisfies properties A, or B, or C,..., then existence and uniqueness will hold. Maybe existence and uniqueness hold in some other cases even when A, B and C,... all fail, but the theorems are silent on that point, and each case must then be treated on its merits. Usually, conditions A, B, C,... relate to some regularity property or other of the function Q.

The proof of existence and uniqueness theorems for differential equations belongs to an area of mathematics which is beyond the scope of this book. Nevertheless, these results have an important bearing on the structure of solution sets of differential equations. For the reader who wants to find out more, I have included a statement of one of the more useful existence and uniqueness results in the following sections, together with the sketch of a proof in a simple case. We shall see that existence and uniqueness of solutions depends on the **differentiability** of $Q(x_0, y)$ with respect to y. The fact that $d(3y^{2/3})/dy$ fails to exist at $y = 0$ leads to the breakdown of uniqueness for the DE $df/dx = 3f^{2/3}$; whereas $df/dx = f^2$, for which $d(y^2)/dy$ exists for all y, has a unique solution subject to any given initial condition.

EXERCISES ON 10.1

1. Solution curves of the DE $x\, df/dx = -f$ are described by the equation $y = f(x)$, where $dy/dx = -y/x$. Select a number of points (x_0, y_0), with $x_0 \neq 0$, and at each point draw a short line segment having slope $-y_0/x_0$. Draw sufficiently many of these line segments so that you can sketch some of the solution curves which connect up some of the segments. (For those who are familiar with them, these are like the 'iron filing' curves which indicate the directions of magnetic field lines.)

By considering the derivative $d(xy)/dx$, find the general solution of the DE, and find the equation of the solution curve which passes through a given point (x_0, y_0), with $x_0 \neq 0$.

2. Draw some of the line segments and sketch 'iron filing' curves, as in problem 1, for the DE $df/dx = -x/f$.

   By considering $d(x^2 + f^2)/dx$, find the general solution of the DE. Find the equation of the unique solution curve passing through a given point (x_0, y_0), with $y_0 \neq 0$.

3. The motion of a point P along the x-axis is described by the function of time $x = x(t)$, where $x(t)$ satisfies the DE $dx/dt = 2\sqrt{x+1}$.

   Use the chain rule to evaluate

$$\frac{d}{dt}\sqrt{x+1}$$

   Hence show that P moves with constant acceleration, and determine $x(t)$ for $t \geq 0$, given that P is at the origin at time $t = 0$.

4. Solve the DE $df/dx = f^2/x^2$, subject to the initial condition $f(1) = 0$. What is the maximal interval of x values for which this solution is valid?

5. Find the general solution of the DE

$$\frac{dy}{dx} = \exp(x + y)$$

   Determine also the particular solution which satisfies the initial condition $y(0) = 0$. What is the maximal interval of definition of this solution?

10.2 Series solutions and iterations

In Section 10.1, we examined a number of examples of first order DEs. Of these, the equation $df/dx = x + f$ is an example of a class of DE described as **linear**, in view of the linear dependence on f of the right-hand side. We shall look more closely at linear first order DEs in Chapter 11. Any linear first order DE can be solved explicitly, in the sense that the determination of the general solution can be reduced to the evaluation of just two integrals. We shall also consider the solution of linear DEs of second order in Chapter 11.

The other DEs treated in Section 10.1 were either separable, or at least could be reduced to separable equations by means of simple transformations. It should be realized, however, that very few first order DEs are either linear, or separable, or can be reduced to a solvable form in a straightforward way. In the great majority of cases, no explicit formula for the solution of a DE $df/dx = Q(x, f)$ can be found. This does not, of course, mean that we simply throw up our hands and abandon such a problem. As I pointed out earlier, even an equation like $df/dx = f$ has no solution (apart from the zero function) if we insist that every solution must be capable of exact evaluation for every value of x, and the fact that we regard this equation as solved, with $f(x) = A\exp(x)$, is very much dependent on the historical fact that the exponential function has been defined, provided with an appropriate notation, and investigated mathematically. Just as the properties of the exponential function can be explored via a series expansion, so it is possible to obtain series solutions for most of the DEs which are of practical interest.

A second possible approach to the first order DE is to look for an **iterative** solution. The method of **iteration**, or successive approximation, can also be used to solve **algebraic** equations. The idea is to set up a recursive formula $x_{n+1} = g(x_n)$, in which each successive approximation x_n to a solution of the algebraic equation is replaced by a succeeding approximation x_{n+1}, in which x_{n+1} is some appropriate function of x_n. Given a suitable starting point x_0 for the iteration, and a well chosen function g, x_{n+1} will be a **closer** approximation than x_n, and in the limit as n tends to infinity x_n will converge to an exact solution of the given algebraic equation. You will find an example of the iterative solution of an algebraic equation in exercise 9 at the end of Chapter 6.

A similar idea can be applied to solve differential rather than algebraic equations. In fact, you have already seen such a method in action in Example 5 of Section 9.2, where an iterative sequence of approximations to the exponential function was obtained (see also Exercises on 9.2). To transform the differential equation into a form suitable for iteration, it is necessary first of all to write down the corresponding integral equation (see Section 9.2). This integral equation can then be iterated, to provide a sequence of approximating solutions to the original differential equation.

I shall consider in turn these two approaches to the solution of first order DEs, the series method and the iterative method. Each method provides a useful addition to our armoury in attacking this type of problem; and iteration leads to something else — a proof of existence and uniqueness for solutions of first order DEs.

The series method of solution is best illustrated by means of an example.

Example 5

Series solution of first order DE: we apply the method to the equation $df/dx = x + f^2$, subject to the initial condition $f(0) = 1$. The idea is to look for a series solution, of the form $f(x) = a_0 + a_1 x + a_2 x^2 + a_3 x^3 + \dots$.

Assuming that such a series solution exists and converges (for $|x| < R$ where R is the radius of convergence; see Section 6.1), the most direct way to determine coefficients is through the Taylor expansion $f(x) = f(0) + xf'(0) + (x^2/2!)f''(0) + (x^3/3!)f'''(0) + \dots$. To evaluate a_n, we need to determine the derivative of f of order n, evaluated at $x = 0$. This can be done by using the DE, and the derivative of the DE, and the derivative of the derivative of the DE, etc. The whole process can be started off by using the initial condition. Thus $f(0) = 1$, from the initial condition.

The successive derivatives of f are:

$$\frac{df}{dx} = x + f^2$$

$$\frac{d^2f}{dx^2} = \frac{d}{dx}(x + f^2) = 1 + 2f\frac{df}{dx}$$

$$\frac{d^3f}{dx^3} = \frac{d}{dx}\left(1 + 2f\frac{df}{dx}\right) = 2f\frac{d^2f}{dx^2} + 2\left(\frac{df}{dx}\right)^2$$

$$\frac{d^4f}{dx^4} = \frac{d}{dx}\left(2f\frac{d^2f}{dx^2} + 2\left(\frac{df}{dx}\right)^2\right) = 2f\frac{d^3f}{dx^3} + 6\frac{df}{dx}\frac{d^2f}{dx^2}$$

$$\frac{d^5f}{dx^5} = \frac{d}{dx}\left(2f\frac{d^3f}{dx^3} + 6\frac{df}{dx}\frac{d^2f}{dx^2}\right) = 2f\frac{d^4f}{dx^4} + 8\frac{df}{dx}\frac{d^3f}{dx^3} + 6\left(\frac{d^2f}{dx^2}\right)^2$$

$$\frac{d^6f}{dx^6} = \frac{d}{dx}\left(2f\frac{d^4f}{dx^4} + 8\frac{df}{dx}\frac{d^3f}{dx^3} + 6\left(\frac{d^2f}{dx^2}\right)^2\right)$$

$$= 2f\frac{d^5f}{dx^5} + 10\frac{df}{dx}\frac{d^4f}{dx^4} + 20\frac{d^2f}{dx^2}\frac{d^3f}{dx^3},$$

... and so on.

I hope you will take the trouble to verify these expressions; they are good examples of the application of the product rule to evaluate higher order derivatives.

Substituting $x = 0$ now gives $f'(0) = 0 + (f(0))^2 = 0 + 1 = 1$; $f''(0) = 1 + 2f(0)f'(0) = 1 + 2.1.1 = 3$; $f'''(0) = 2f(0)f''(0) + 2(f'(0))^2 = 2.1.3 + 2.1^2 = 8$; and the remaining derivatives, up to the sixth, at $x = 0$, are evaluated in a similar way, with each successive derivative at $x = 0$ depending on the previous evaluation of lower order derivatives. The values of f, df/dx, and higher order derivatives up to the sixth, at $x = 0$, are given by

$$f = 1, \quad \frac{df}{dx} = 1, \quad \frac{d^2f}{dx^2} = 3, \quad \frac{d^3f}{dx^3} = 8, \quad \frac{d^4f}{dx^4} = 34, \quad \frac{d^5f}{dx^5} = 186, \quad \frac{d^6f}{dx^6} = 1192$$

Substituting these values into the Taylor expansion for the solution f about $x = 0$ now gives

$$f(x) = 1 + x + \frac{3x^2}{2} + \frac{4x^3}{3} + \frac{17x^4}{12} + \frac{31x^5}{20} + \frac{149x^6}{90} + \dots \tag{10.1}$$

In principle, the coefficients of the series may be determined up to any required order, and the sum of the series should then give the required solution $f(x)$ of the DE. But we cannot expect the power series, of which the terms up to order x^6 are given in equation (10.1), to converge for all values of x. This would imply that the solution $f(x)$ was defined for all $x \in \mathbb{R}$. In fact, $f(x)$ must tend to infinity for some value of x between 0 and 1. To see why this is so, we must return to the DE and initial condition. Thus $df/dx = x + f^2$; $f(0) = 1$.

It is clear from the DE that df/dx is positive for x positive, and hence that f is an increasing function for $x > 0$. Since $f(0) = 1$, it follows that $f(x) \geq 1$ for $x \geq 0$. We can expect that, as x increases, the term f^2 on the right-hand side of the DE will be increasingly important, and to take this term into account we shall treat the DE rather similarly to the equation $df/dx = f^2$, for which we should have to consider the derivative $d(-1/f)/dx$. In this case, then, for any $x > 0$ in the domain of f, we have, by the chain rule

$$\frac{d}{dx}\left(-\frac{1}{f}\right) = \frac{1}{f^2}\frac{df}{dx} = \frac{1}{f^2}(x + f^2) = \frac{x}{f^2} + 1$$

Since $f(x) \geq 1$, this tells us that (for $x > 0$ in the domain of f), $d(-1/f)/dx$ lies between 1 and $1 + x$.

Having bounds for the **derivative** of a function leads, through integration, to bounds for the function itself. Integrating with respect to t from 0 to x the inequality

$$1 \leq \frac{d}{dt}\left(-\frac{1}{f(t)}\right) \leq 1 + t$$

now gives

$$x \leq 1 - \frac{1}{f(x)} \leq x + \frac{x^2}{2}$$

for all $x > 0$ at which the solution $f(x)$ is defined. For such f, the inequality $x \leq 1 - (1/f(x))$ leads to the estimate $f(x) \geq 1/(1 - x)$. (What we have shown is that, with $f(0) = 1$, the solution of $df/dx = x + f^2$ must be larger than the solution of $df/dx = f^2$.)

Now any function which is increasing for $x > 0$, and which satisfies an inequality $f(x) \geq 1/(1 - x)$ must diverge (i.e. tend to infinity) as x approaches some value between 0 and 1. The solution $f(x)$ cannot be defined for any $x \geq 1$. We can make things more precise by looking at the other half of our inequality, $1 - (1/f(x)) \leq x + (x^2/2)$. For x in the interval $0 \leq x \leq 0.7$, this implies $1 - (1/f(x)) \leq 0.945$, or $f(x) \leq 1/0.055$. The precise inequality does not matter; what matters is that the solution $f(x)$ must be **bounded** ($f(x) < 20$, say) at least for x in the interval $0 \leq x \leq 0.7$ — so $f(x)$ cannot diverge **before** we reach the point $x = 0.7$.

We have now shown, from the DE and initial condition, that the solution diverges at some point $x = R$ in the interval $(0.7, 1]$. I have used the symbol R here, because, for a power series having radius of convergence R, we will have convergence for $|x| < R$ and divergence for $|x| > R$. The smallest positive value of x at which the power series in equation (10.1) diverges may be identified with the value of x at which the **solution** $f(x)$ of the DE diverges, and is the radius of convergence of the power series. It may be verified, for $|x| < R$, that the series in equation (10.1) does indeed converge to the required solution of the DE. Our analysis of the solution of the DE for positive values of x has led us to estimate $0.7 < R \leq 1$ for the radius of convergence of the power series. A more direct method of estimating the radius of convergence is by bounding the coefficients of the series. In problem 1 of Exercises on 10.2, it is shown that the coefficient of x^n in the series is bounded by $(3/2)^{n/2}$, and this leads to the improved estimate $0.8 < R \leq 1$.

The series method for first order DEs cannot be recommended as a computationally efficient method for use with high speed computers. It does, however, provide a reasonably quick method for investigating the local behaviour of solutions. If the initial condition is at $x = x_0$, then of course the method has to be modified by using the Taylor expansion about this point, a series in powers of $(x - x_0)$ rather than x. In the present example, the series as in equation (10.1), up to order x^6, may be used to obtain an approximate value for the solution for a range of values of x. At $x = 0.4$, for example, the series to this order gives $f(x) = 1.784...$, whereas the exact solution at this point is $f(x) = 1.789....$ This is an accuracy of better than a third of one per-cent, and one per-cent accuracy is maintained up to about $x = 0.48$. Similar accuracy is obtained for negative values of x, though the series method has the disadvantage that the series converges only for $|x| < R$, whereas the solution $f(x)$ exists for all x in the range $-\infty < x < R$. (For a qualitative investigation of the solution of this DE for $x < 0$, see problem 2 of Exercises on 10.2. The series solution in powers of $(x + 0.4)$, starting with the correct initial condition at $x = -0.4$, will, however, converge for all x in the interval $(-0.8 - R, R)$, which has -0.4 as its midpoint, and extends at least to -1.6 at the left-hand endpoint). Quite apart from its numerical accuracy, the series method has considerable value as an analytical tool in the investigation of a variety of first order DEs.

The second approach to the solution of first order linear DEs that I shall describe is the method of iteration. To begin with, the DE, together with the initial condition, must be converted to an **integral equation**. We start with the first order DE in its most general form:

$$\frac{df}{dx} = Q(x, f)$$

subject to an initial condition $f(x_0) = y_0$.

We can use the fundamental theorem of calculus to write down an antiderivative of the right-hand side of the DE. We have, in fact,

$$\frac{d}{dx} \int_{x_0}^{x} Q(t, f(t)) dt = Q(x, f(x))$$

so that the DE implies

$$\frac{d}{dx} \left\{ f(x) - \int_{x_0}^{x} Q(t, f(t)) dt \right\} = 0 \tag{10.2}$$

I have assumed, here, that the integrand $Q(t, f(t))$ defines a continuous function of t on the interval $x_0 \leq t \leq x$, or on the interval $x \leq t \leq x_0$ if x is smaller than x_0; since f is necessarily a continuous function, this assumption will hold provided $Q(x, y)$ depends continuously on x and y.

Equation (10.2) holds on any interval on which f is defined and satisfies the DE. Such an interval must, of course, contain the point x_0 at which the initial condition is given. It follows from equation (10.2) that

$$f(x) - \int_{x_0}^{x} Q(t, f(t)) dt = c$$

where c is constant on the interval in question. Setting $x = x_0$, in which case the integral is zero, we have $f(x_0) = c$. Hence $c = y_0$ from the initial condition, and we obtain the so-called Volterra integral equation

$$f(x) = y_0 + \int_{x_0}^{x} Q(t,f(t))dt. \tag{10.3}$$

Any solution f of the DE, subject to the initial condition $f(x_0) = y_0$, will satisfy the integral equation and, conversely, any solution f of the integral equation will satisfy both the DE and the initial condition. We can say, then, that the integral equation (10.3) is equivalent to DE +initial condition. Equation (10.3) does not of itself solve our problem because in order to use this equation to determine $f(x)$, for given initial condition and given function Q, we should need to substitute for $f(t)$ into the integrand on the right-hand side; but $f(t)$ is unknown, until we have determined the function f!

What we **can** do is set up an iterative sequence of functions $f_0, f_1, f_2, f_3,...$ which successively approximate to the required solution. A suitable starting point for the iteration might be $f_0(x) = y_0$, the constant function, after which each succeeding function in the sequence is defined in terms of the previous function by the iterative formula

$$f_{n+1}(x) = y_0 + \int_{x_0}^{x} Q(t,f_n(t))dt \tag{10.4}$$

The hope is that f_{n+1} is always a better approximation to the exact solution than f_n, and that as n tends to infinity the sequence of functions $\{f_n(x)\}$ converges to an exact solution $f(x)$ for each x in the interval of definition. Let us see how this iteration works for a simple example. For comparison purposes, we cannot do better than consider the same DE and initial condition that we have already treated, in Example 5, by the series method.

Example 6

Iterative solution of first order DE: we consider once more the DE. $df/dx = x + f^2$, subject to the initial condition $f(0) = 1$. Then

$$\frac{d}{dx}\left\{\frac{x^2}{2} + \int_0^x (f(t))^2 dt\right\} = x + (f(x))^2$$

and the DE implies

$$\frac{d}{dx}\left\{f(x) - \frac{x^2}{2} - \int_0^x (f(t))^2 dt\right\} = 0$$

Integrating and using the initial condition $f(0) = 1$ now leads to the integral equation

$$f(x) = 1 + \frac{x^2}{2} + \int_0^x (f(t))^2 dt \tag{10.5}$$

We can now iterate equation (10.5) by taking the constant function $f_0(x) = 1$ as the starting point (an alternative would be $f_0(x) = 1 + x^2/2$) and defining successive functions in the iteration by

$$f_{n+1}(x) = 1 + \frac{x^2}{2} + \int_0^x (f_n(t))^2 dt \tag{10.6}$$

Thus

$$f_0(x) = 1, f_1(x) = 1 + \frac{x^2}{2} + \int_0^x 1 dt = 1 + x + \frac{x^2}{2},$$

$$f_2(x) = 1 + \frac{x^2}{2} + \int_0^x \left(1 + t + \frac{t^2}{2}\right)^2 dt = 1 + x + \frac{3x^2}{2} + \frac{2x^3}{3} + \frac{x^4}{4} + \frac{x^5}{20}$$

and so on. The next step in the iteration is to substitute $f_2(t) = 1 + t + (3t^2/2) + (2t^3/3) + (t^4/4) + (t^5/20)$ into the integrand on the right-hand side of equation (10.6). This results in a polynomial of degree 11 for $f_3(x)$, of which the first few terms are

$$f_3(x) = 1 + x + \frac{3x^2}{2} + \frac{4x^3}{3} + \frac{13x^4}{12} + \dots$$

For $f_4(x)$ we find, in a similar way on substituting for $f_3(t)$,

$$f_4(x) = 1 + x + \frac{3x^2}{2} + \frac{4x^3}{3} + \frac{17x^4}{12} + \dots$$

Notice how already, at the fourth iteration, the iterative approximation $f_4(x)$ agrees up to power x^4 with the series solution obtained previously in Example 5.

It would be nice to prove that the iteration is convergent, in the sense that $f_n(x)$ converges to a limit as n tends to infinity, at least in some interval of x values which includes $x = x_0$. To do so, a first step is to show that the functions f_n remain **bounded**. Provided the interval is short enough, this first step can usually be achieved by the method of induction. For example, restrict x to the interval $-0.2 \le x \le 0.2$, and assume inductively that $|f_n(x)| \le 2$ on this interval. Certainly the inductive hypothesis is satisfied with $n = 0$, since we have $f_0(x) = 1$ in that case. Moreover, on the right-hand side of equation (10.6) the integral cannot exceed $\int_0^{0.2} (2)^2 dt$ if x is positive (if x is negative, then minus the integral cannot exceed this value); so the largest possible value of $f_{n+1}(x)$ on the given interval is $1 + [(0.2)^2/2] + 0.2 \times 2^2 = 1.82$ (if x is negative, an even smaller bound for $|f_{n+1}(x)|$ on the interval is obtained).

Since $1.82 < 2$, we have shown that if $|f_n(x)| \le 2$ on the interval $[-0.2, 0.2]$, then $|f_{n+1}(x)| \le 2$ on the same interval. Since $|f_0(x)| \le 2$, this is sufficient to show that $|f_n(x)| \le 2$ on the interval, for **all** values of n. In particular, all functions of the iterative sequence remain bounded on this interval. (In fact, this result may be strengthened to show that the f_ns are bounded on an interval at least as large as $(-0.8, 0.8)$: see problem 3 of Exercises on 10.2.)

Having shown that the functions of the iterative sequence are bounded, it is possible to give the sketch of a proof of convergence of the iteration. The basic idea is to use equation (10.6) to compare consecutive f_n, f_{n+1} of the iterative sequence. Equation (10.6) and the corresponding equation with $n - 1$ instead of n tell us that

$$\left. \begin{array}{l} f_{n+1}(x) = 1 + \dfrac{x^2}{2} + \displaystyle\int_0^x (f_n(t))^2 dt \\[3mm] f_n(x) = 1 + \dfrac{x^2}{2} + \displaystyle\int_0^x (f_{n-1}(t))^2 dt \end{array} \right\}$$

Subtracting these two equations, we have

$$f_{n+1}(x) - f_n(x) = \int_0^x \{(f_n(t))^2 - (f_{n-1}(t))^2\} \, dt \tag{10.7}$$

The integrand on the right-hand side can be factorized as $(f_n(t) + f_{n-1}(t))(f_n(t) - f_{n-1}(t))$. If, then, we restrict x to lie in a finite interval containing $x = 0$ on which the f_ns are bounded, say $|f_n| \leq k$ for all n, the factor $f_n(t) + f_{n-1}(t)$ cannot exceed $2k$ in absolute value. Hence $|(f_n(t))^2 - (f_{n-1}(t))^2| \leq 2k|f_n(t) - f_{n-1}(t)|$, and it follows from equation (10.7) that

$$|f_{n+1}(x) - f_n(x)| \leq 2k \left| \int_0^x |f_n(t) - f_{n-1}(t)| \, dt \right| \tag{10.8}$$

With $f_0(t) = 1$ and $f_1(t) = 1 + t + (t^2/2)$, we know that $|f_1(t) - f_0(t)|$ is bounded on the interval in question; say $|f_1(t) - f_0(t)| \leq C$.

Substituting this estimate into the right-hand side of equation (10.8), with $n = 1$, we have $|f_2(x) - f_1(x)| \leq 2k |\int_0^x C \, dt| = 2kC|x|$. Hence $|f_2(t) - f_1(t)| \leq 2kC|t|$. This in turn can be substituted into the right-hand side of equation (10.8), with $n = 2$, to give

$$|f_3(x) - f_2(x)| \leq 2k \left| \int_0^x 2kCt \, dt \right| = (2k)^2 C x^2/2$$

Proceeding in this way, we have

$$|f_4(x) - f_3(x)| \leq (2k)^3 C|x|^3/3!$$

and, quite generally,

$$|f_{n+1}(x) - f_n(x)| \leq (2k|x|)^n C/n! \tag{10.9}$$

Now $f_n(x)$ may be written as a sum

$$f_n(x) = f_0(x) + \{f_1(x) - f_0(x)\} + \{f_2(x) - f_1(x)\} + \ldots + \{f_n(x) - f_{n-1}(x)\}$$

Hence the convergence of $f_n(x)$ to a limit as n tends to infinity will depend on the convergence of the infinite series $\sum_0^\infty \{f_{n+1}(x) - f_n(x)\}$; but by equation (10.9), the terms of this series are bounded in absolute value by terms of the exponential series $\sum_0^\infty (2k|x|)^n . C/n!$ $(= C \exp 2k|x|)$. By comparison of the two series, the convergence of the exponential series guarantees convergence of the series of differences of the functions f_n, so $f_n(x)$ converges to a limit, for all x in the interval on which these functions are bounded, and we have shown convergence of the iteration!

One final step is necessary, in order to show that the function $f(x)$ which is the limit of the iteration, $f(x) = \lim_{n \to \infty} f_n(x)$, does indeed satisfy the DE together with the initial condition. To do this, we have simply to take the limit $n \to \infty$ in equation (10.6). The left-hand side of the equation converges in that case to $f(x)$, whereas the right-hand side converges to $1 + (x^2/2) + \int_0^x (f(t))^2 \, dt$. (Strictly, we need to show that $\lim_{n \to \infty} \int_0^x (f_n(t))^2 \, dt = \int_0^x (f(t))^2 \, dt$ where $f(t) = \lim_{n \to \infty} f_n(t)$: such a step can in fact be justified, but requires the idea of uniform convergence, for which the reader will have to consult a more advanced text of analysis.)

Equating the respective limits of the right-hand side and left-hand side of equation (10.6), we now see that the limiting function $f(x)$ of the iteration does indeed satisfy the integral equation, and hence the DE as well as the initial condition. The method of iteration for solution of first order DEs does really work!

EXERCISES ON 10.2

[These problems all relate to the solution of the DE $df/dx = x + f^2$, subject to the initial condition $f(0) = 1$.]

1. (a) Show that the solution f satisfies the equation $d^{n+1}f/dx^{n+1} = d^n/dx^n f^2$, for all $n \geq 2$.

    (b) Assuming that the nth derivative on the right-hand side is given (according to Leibniz's formula: see problem 2 of Exercises on 6.1) by

    $$\frac{d^n f^2}{dx^n} \equiv D^n f^2 = f.D^n f + {}^n C_1 Df.D^{n-1}f + {}^n C_2 D^2 f.D^{n-2}f + \dots + D^n f.f$$

    verify that if $|f| \leq 1$, $|Df| \leq c$, $|D^2 f| \leq c^2.2!$, $|D^3 f| \leq c^3.3!$, $\dots |D^n f| \leq c^n.n!$, where c is a constant ≥ 1, then $|D^{n+1}f| = |D^n f^2| \leq c^{n+1}(n+1)!$

    (c) Taking $c = \sqrt{\frac{3}{2}}$ in (b), verify the inequality $|D^n f| \leq (\frac{3}{2})^{n/2}.n!$ in the cases $n = 1, 2, 3$ at $x = 0$; prove by induction that $|D^n f| \leq (\frac{3}{2})^{n/2}.n!$ at $x = 0$, for all $n \geq 1$.

    (d) Show that the coefficient of x^n in the Taylor series (equation (10.1)) of $f(x)$ about $x = 0$ is bounded by $(\frac{3}{2})^{n/2}$. Use the ratio test to deduce that the series is convergent for $|x| < \sqrt{\frac{2}{3}}$. What does this tell you about the radius of convergence of the series?

2. A function $g(x)$ is defined for $x \geq 0$ by $g(x) = f(-x)$, where f is the solution of the DE, subject to the given initial condition. Show that the solution curve $y = g(x)$ is described by the equation $dy/dx = x - y^2$, and passes through the point $x = 0$, $y = 1$.

    (a) For the values $c = -1, -\frac{1}{2}, 0, \frac{1}{2}, 1$, sketch **on a single graph** the curves $x - y^2 = c$, intersected for $y > 0$ by short line segments having gradient c. Your curve with $c = 0$ should look as in Fig 10.2. Since the solution curve of $dy/dx = x - y^2$ intersects $x - y^2 = 0$ with slope $dy/dx = 0$, each line segment represents a possible **intersection** of the solution curve with $x - y^2 = 0$, and similarly for intersections with $x - y^2 = c$ for the other values of c.

    Note that the solution curve starts by cutting $x - y^2 = -1$ at the point $(0,1)$, with slope -1. Use your completed diagram to provide a rough sketch of the solution curve $y = g(x)$ for $x > 0$.

    (b) Use the equation $dy/dx = x - y^2$ to deduce the following properties of the solution curve $y = g(x)$:

    (i) $y > 0$ for all $x > 0$ (if $x = x_1$ were the smallest value of x at which $y = 0$, show that $y'(x_1) > 0$, and explain why this cannot happen);

    (ii) $d/dx\,(x - y^2) \geq 1$ provided $x - y^2 \leq 0$, hence there is a point a between 0 and 1 such that $x - y^2$ increases between 0 and a, and $x - y^2 = 0$ at $x = a$;

    (iii) $x - y^2 > 0$ for $x > a$, and $x - y^2 \to 0$ as $x \to \infty$.

    [Hence the solution of $df/dx = x + f^2$, subject to $f(0) = 1$, satisfies $f(x) - \sqrt{-x} \to 0$ as $x \to -\infty$.]

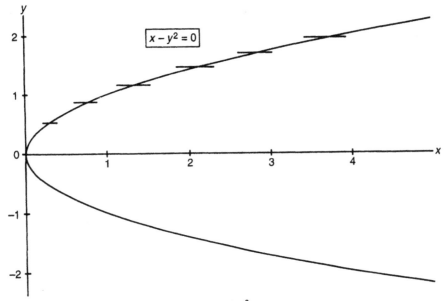

Fig 10.2 $x - y^2 = 0$.

3. An iterative sequence of approximate solutions of the DE $df/dx = x + f^2$, subject to the initial condition $f(0) = 1$, is defined by

$$f_{n+1}(x) = 1 + \frac{x^2}{2} + \int_0^x (f_n(t))^2 dt$$

For x in the interval $0 \leq x < 1/c$, where c is a constant greater than unity, show that if $f_n(x) \leq (1 - cx)^{-1}$, then

$$f_{n+1}(x) \leq 1 + \frac{x^2}{2} + \int_0^x (1 - ct)^{-2} dt = 1 + \frac{x^2}{2} + \frac{x}{1 - cx}$$

Deduce that $f_{n+1}(x) \leq (1 - cx)^{-1}$ provided $cx^2 - x + 2(c - 1) > 0$, and verify that this inequality is satisfied in the case $c = 1.12$. Hence prove by induction that there is a positive constant k such that $0 \leq f_n(x) \leq k$ for all $n > 0$ and x in the interval $0 \leq x \leq 0.89$.

10.3 How many solutions?

I hope that our discussion of examples of first order DEs will have given you a good idea of the variety of methods available, and what to expect in the way of solutions. Our expectation is that an equation of the form $df/dx = Q(x, f)$ will have one, and only one, solution satisfying a given initial condition $f(x_0) = y_0$. This solution need not be defined for **all** values of x, but we expect that at least there will be an interval (a, b), containing x_0, such that the solution will exist on this interval. As we reach the edges $x = a$, $x = b$ of this interval, the solution may start to diverge, or wildly oscillate, or become uncontrollable in other ways; but if we stay within the interval $a < x < b$, all should be well, and our solution should be well defined.

We can summarize all this by saying that we can expect a unique solution, subject to $f(x_0) = y_0$, in a **neighbourhood** of x_0, that is in a sufficiently small interval containing x_0. Graphically, this means that we expect to find a unique solution curve $y = f(x)$, satisfying $dy/dx = Q(x, y)$, passing through the point (x_0, y_0) and taking us at least a short distance away from this point (to the right **and** to the left, if we are interested in solving the equation both for $x > x_0$ and for $x < x_0$).

How far are our expectations with respect to existence and uniqueness of solutions born out in practice? As far as the **existence** of the solution near to $x = x_0$ is concerned, in most cases this can be shown using an argument based on iteration, as in the case of the DE $df/dx = x + f^2$, with $f(0) = 1$, discussed in Section 10.2. For such an argument to work, it is first of all necessary to obtain a bound $|f_n(x)| \leq k$ for all of the functions in the iterative sequence. This can easily be done, as in Section 10.2, by induction, if we assume that $Q(x, y)$ is itself a bounded function of x and y, for (x, y) close enough to (x_0, y_0).

Having shown that the f_ns are bounded, we can again prove convergence of the iteration to a solution of the integral equation following the same argument as in Section 10.2. For this, an examination of our previous proof of this result in the case of the DE $df/dx = x + f^2$ will show that this depended on an inequality of the form $|Q(x, y_1) - Q(x, y_2)| \leq \text{const}|y_1 - y_2|$, which again we assume to hold for (x, y_1) and (x, y_2) close enough to (x_0, y_0). In that case we can obtain an iterative solution of the integral equation, which we know is a solution also of the DE + initial condition.

What about the question of **uniqueness** of solution, subject to a given initial condition? This too can be shown, assuming an inequality of the form $|Q(x, y_1) - Q(x, y_2)| \leq \text{const}|y_1 - y_2|$. It is an interesting exercise to follow through the proof, which I have included in problem 1 of Exercises on 10.3. So quite generally, for a wide class of equations, everything that we had hoped for with regard to existence and uniqueness does turn out to be true! The result can be summarized in the following theorem.

● Theorem 1. Existence/Uniqueness for Solutions of First Order DEs

Let $Q(x, y)$ be a continuous function of x and y in some neighbourhood of (x_0, y_0). (By a neighbourhood of (x_0, y_0), I mean that the (x, y) values must be sufficiently close to (x_0, y_0) in the sense that $|x - x_0| < \text{const}$, $|y - y_0| < \text{const}$.) Suppose that, for (x, y_1) and (x, y_2) in this neighbourhood of (x_0, y_0), the function $Q(x, y)$ satisfies an estimate of the form

$$|Q(x, y_1) - Q(x, y_2)| \leq C|y_1 - y_2| \tag{10.10}$$

for some positive constant C. (This estimate also implies that $Q(x, y)$ is bounded for (x, y) in the given neighbourhood.) Then, for x sufficiently close to x_0, there exists one and only one solution $f(x)$ of the DE $df/dx = Q(x, f)$, subject to the initial condition $f(x_0) = y_0$.

Theorem 1 is neither the most general nor the strongest existence and uniqueness theorem that can be found. However, it does give at least an indication of the kind of function that Q must be in order to guarantee a single solution for each initial condition. In fact, the condition on Q can be simplified even further. Provided $Q(x_0, y)$ is a differentiable function of y, with $|(d/dy) Q(x_0, y)| \leq$ const for all x_0 on some interval, an estimate as in equation (10.10) will hold for x in that interval, and follows from the Mean Value Theorem; so existence and uniqueness of solutions will always hold if Q has a bounded derivative with respect to y.

We have dealt so far with the case of **first order** DEs. Let us now consider what happens in the case of second and higher order.

● *Definition 2*

A second order DE for a function $f(x)$ is an equation of the form $d^2 f / dx^2 = Q(x, f, df/dx)$ (or an equation reducible to this form), where Q is a given function of $x, f, df/dx$. Examples are:

$$\frac{d^2 f}{dx^2} = x^2 + f^2 + \left(\frac{df}{dx}\right)^2, \; \frac{d^2 f}{dx^2} = xf^2, \; \frac{d^2 f}{dx^2} = \cos x, \; (1 + x^2)\frac{d^2 f}{dx^2} + x\frac{df}{dx} + f = 0$$

A function f is said to be a solution of the DE on an interval $a < x < b$ if

$$\frac{d^2}{dx^2} f(x) = Q\left(x, f(x), \frac{df(x)}{dx}\right)$$

for all x in the interval. The function Q is assumed to have continuous dependence on $x, f, df/dx$ as x varies over the interval in question. A DE of order n with $n > 2$ is defined in a similar way, and expresses $d^n f / dx^n$ in terms of $x, f, df/dx$ and higher order derivatives up to order $(n - 1)$.

Any DE of order 2 may be written as two **simultaneous** equations involving only first derivatives. To see this in the case of the second order DE $d^2 f / dx^2 = x^2 + f^2 + (df/dx)^2$, define two functions $y_1(x)$, $y_2(x)$ by $y_1 = f$, $y_2 = df/dx$. Then

$$\frac{dy_1}{dx} = \frac{df}{dx} = y_2$$

and

$$\frac{dy_2}{dx} = \frac{d^2 f}{dx^2} = x^2 + y_1^2 + y_2^2$$

We have reduced the single second order DE to two simultaneous first order equations

$$\left. \begin{array}{l} \dfrac{dy_1}{dx} = y_2 \\[2mm] \dfrac{dy_2}{dx} = x + y_1^2 + y_2^2 \end{array} \right\}$$

Let us go a step further, by defining a **two-component** function of x by

$$y(x) = \begin{pmatrix} y_1(x) \\ y_2(x) \end{pmatrix}$$

as well as a **two-component** function of x and y by

$$Q(x,y) = \begin{pmatrix} y_2 \\ x + y_1^2 + y_2^2 \end{pmatrix}$$

Taking dy/dx to mean $\left(\frac{dy_1/dx}{dy_2/dx}\right)$, our first order equation can be written as a single first order DE for the two-component function y, namely

$$\frac{dy}{dx} = Q(x,y) \tag{10.11}$$

Equation (10.11) looks familiar. We have converted the second order DE into a form which looks just like a first order equation, except that we have to deal with a two-component function, $y(x)$, or **pair** of functions, rather than a single function $y(x)$.

The same transformation of a second order DE into a pair of first order equations, and thence into a single first order lookalike, can be carried out in the general case. With $y_1 = f$, $y_2 = df/dx$, the second order DE $d^2f/dx^2 = Q(x,f,df/dx)$ becomes the first order equations

$$\left.\begin{array}{l} \dfrac{dy_1}{dx} = y_2 \\[2mm] \dfrac{dy_2}{dx} = Q(x,y_1,y_2) \end{array}\right\}$$

and may be written as in equation (10.11), with now

$$Q(x,y) = \begin{pmatrix} y_2 \\ Q(x,y_1,y_2) \end{pmatrix}$$

Equation (10.11) is not only notationally convenient and suggestive, but carries with it considerable mathematical implications. Starting from equation (10.11), we can follow through a proof almost identical to the existence and uniqueness proof of Theorem 1 to obtain the following result.

The second order differential equation $d^2f/dx^2 = Q(x,f,df/dx)$ has one and only one solution (for x sufficiently close to x_0), subject to the initial condition $y(x_0) = y_0$. Here $y(x_0)$ has two components, $y_1(x_0)$ and $y_2(x_0)$. In terms of the solution f, these two components are the values of f and df/dx at $x = x_0$, so y_0 must have two components as well. If we call these a and b, then the initial condition $y(x_0) = y_0$ is really **two** initial conditions, namely $f = a$ and $df/dx = b$ at $x = x_0$. We are led, then, to the following existence/uniqueness result for second order DEs: for x sufficiently close to x_0, there exists one and only one solution of the DE $d^2f/dx^2 = Q(x,f,df/dx)$ subject to the **pair** of initial conditions $f = a$, $df/dx = b$ at $x = x_0$.

(Such a result, must, of course, require bounds to be placed on the function Q. In fact, it is only necessary to replace the estimate equation (10.10) in Theorem 1 by a corresponding estimate for the two-component function Q, namely $|Q(x,y_1) - Q(x,y_2)| \leq C|y_1 - y_2|$, where $|y|$ is defined for any two-component y by $\left|\begin{pmatrix} \alpha \\ \beta \end{pmatrix}\right| = \sqrt{\alpha^2 + \beta^2}$.)

A similar result holds for DEs of any order n. Here a solution exists and is uniquely defined by n initial conditions; these conditions specify the values of f, df/dx, ..., $d^{n-1}f/dx^{n-1}$ at some initial point $x = x_0$. We should not really be surprised at this result. Given the values of f, df/dx, and all higher order derivatives up to order $(n-1)$ at $x = x_0$, the DE itself will tell us the nth derivative at this point. The derivative of the DE will then allow us to calculate the $(n+1)$th derivative, and repeated differentiation will lead in principle to the value of **all** higher order derivatives at $x = x_0$. A Taylor series expansion about $x = x_0$ will then give us the required solution, which may be expected to converge at least near x_0. The following examples, which extend the series method of Section 10.2 to DEs of second order, put this idea into action.

● *Example 7*

We apply the series method to the second order DE $d^2f/dx^2 = f^2$, subject to the two initial conditions $f(0) = 0$, $f'(0) = 1$. Then I shall explain a method which leads to an 'exact' solution of this DE.

The initial conditions prescribe the values of f and f' at $x = 0$. To obtain f'' and higher order derivatives, we use the DE and derivatives of the DE, making use of the product rule for differentiation. We go up to the derivative of order 7.

$$\left. \begin{aligned}
\text{Thus } f'' &= f^2 \\
f''' &= \frac{d}{dx}(f^2) = 2ff' \\
f'''' &= 2ff'' + 2(f')^2 \\
f''''' &= 2ff''' + 6f'f'' \\
f'''''' &= 2ff'''' + 8f'f''' + 6(f'')^2 \\
f''''''' &= 2ff''''' + 10f'f'''' + 20f''f'''
\end{aligned} \right\}$$

At $x = 0$, substituting for f and f', we can successively evaluate the other derivatives, to give $f = 0$, $f' = 1$, $f'' = 0$, $f''' = 0$, $f'''' = 2$, $f''''' = 0$, $f'''''' = 0$, $f''''''' = 20$.

Up to order x^7, the Taylor series for $f(x)$ has just three terms, and is given by

$$f(x) = x + \frac{x^4}{12} + \frac{x^7}{252} + \dots$$

Another method of solution, which extends to any DE of the form $d^2f/dx^2 = Q(f)$, is as follows. Multiply both sides of the equation $d^2f/dx^2 = f^2$ by $2df/dx$, to obtain

$$2\frac{df}{dx}\frac{d^2f}{dx^2} = 2f^2\frac{df}{dx}$$

The left-hand side is now $(d/dx)(df/dx)^2$, and the right-hand side is $(d/dx)(\tfrac{2}{3}f^3)$. Hence

$$\frac{d}{dx}\left\{\left(\frac{df}{dx}\right)^2 - \frac{2}{3}f^3\right\} = 0$$

from which it follows that $(df/dx)^2 - \frac{2}{3}f^3 = c$, where c is a constant. When $x = 0$, $f = 0$ and $df/dx = 1$, so we must have $c = 1$. Hence

$$\left(\frac{df}{dx}\right)^2 = 1 + \frac{2}{3}f^3$$

This is now a **first order** DE, which on writing $y = f(x)$ becomes $dy/dx = \sqrt{1 + \frac{2}{3}y^3}$; we take the positive square root since $dy/dx > 0$ at $x = 0$.

Having solved first order equations of this kind in Section 10.1, we follow the same method and consider an antiderivative of $1/\sqrt{1 + \frac{2}{3}y^3}$, regarded as a function of y. Such an antiderivative is $\int_0^y (1/\sqrt{1 + \frac{2}{3}t^3})\, dt$, and by the chain rule we have

$$\frac{d}{dx}\int_0^y \frac{1}{\sqrt{1 + \frac{2}{3}t^3}}\, dt = \left(\frac{d}{dy}\int_0^y \frac{1}{\sqrt{1 + \frac{2}{3}t^3}}\, dt\right) \cdot \frac{dy}{dx}$$

$$= \frac{1}{\sqrt{1 + \frac{2}{3}y^3}} \cdot \frac{dy}{dx} = 1$$

On integration, this gives $\int_0^y (1/\sqrt{1 + \frac{2}{3}t^3})\, dt = x + C$, where C is another constant. However, the first initial condition tells us that $y = 0$ when $x = 0$, and it follows that $C = 0$.

Hence x and y are related by the equation

$$\int_0^y \frac{1}{\sqrt{1 + \frac{2}{3}t^3}}\, dt = x$$

Although this equation expresses x as a function of y rather than y as a function of x, such information is just as useful in describing the solution curve.

Since the integral on the left-hand side cannot exceed $\beta = \int_0^\infty (1/\sqrt{1 + \frac{2}{3}t^3})\, dt$ (which is a convergent improper integral; see Section 9.3), x cannot exceed this value either, and it is easily verified that y tends to infinity as x approaches the value of this integral. (If you are not sure about this, sketch the graph $y = \int_0^x (1/\sqrt{1 + \frac{2}{3}t^3})\, dt$, and then invert the graph as described in Section 3.2.) As x moves into negative values, y becomes negative too, until we reach the value $y = -(\frac{3}{2})^{\frac{1}{3}}$. This happens at $x = -\alpha$, where $\alpha = -\int_0^{-(\frac{3}{2})^{\frac{1}{3}}} (1/\sqrt{1 + \frac{2}{3}t^3})\, dt$. Making a change of variable $t = -s$, we see that the value of α is given by the improper integral $\alpha = \int_0^{(\frac{3}{2})^{\frac{1}{3}}} (1/\sqrt{1 - \frac{2}{3}s^3})\, ds$. At $x = -\alpha$, the first order DE for y tells us that $dy/dx = 0$, and the second order DE for f tells us that $d^2y/dx^2 = y^2 = (\frac{3}{2})^{\frac{2}{3}}$.

For $x < -\alpha$, dy/dx must therefore be negative, and our original first order DE must be modified to $\frac{dy}{dx} = -\sqrt{1 + \frac{2}{3}y^3}$. In problem 2 of Exercises on 10.3, you are asked to obtain the solution $x = -2\alpha - \int_0^y (1/\sqrt{1 + \frac{2}{3}t^3})\, dt$, which is valid for $-2\alpha - \beta < x < -\alpha$. The solution curve, which has asymptotes at both $x = -2\alpha - \beta$ and $x = \beta$, is in fact symmetric about $x = -\alpha$. Its graph is sketched in Fig 10.3.

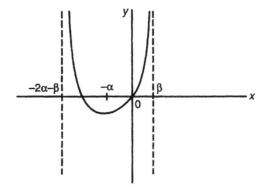

Fig 10.3 Solution curves of $d^2 f/dx^2 = f^2$.

⚙ *Example 8*

A rather simpler example than the previous one illustrates how the series method often leads to dramatic simplifications in the case of **linear** DEs. We shall examine linear DEs in Chapter 11, in particular looking at the structure of the solution set and finding methods for their solution.

Consider the DE $d^2 f/dx^2 + f = 0$, subject to the initial conditions $f(0) = 1$, $f'(0) = 0$. You can probably figure out the solution already, but in case you cannot, note that, at $x = 0$, $f = 1$, $df/dx = 0$, $d^2 f/dx^2 = -f = -1$, $d^3 f/dx^3 = -df/dx = 0$, $d^4 f/dx^4 = -d^2 f/dx^2 = 1$, and so on. This leads to the Taylor expansion, about $x = 0$

$$f(x) = 1 - \frac{x^2}{2!} + \frac{x^4}{4!} - \dots$$

which after a few more terms you can convince yourself is the trigonometric series for $\cos x$. So the exponential function is the solution of a very simple first order DE, with an initial condition $(df/dx = f, f(0) = 1)$ and the cosine function (as well as the sine function) is the solution of a very simple **second order** DE, again with initial conditions (in the case of $\sin x$, $d^2 f/dx^2 = -f$, $f(0) = 0$, $f'(0) = 1$). Just about **all** the functions which we meet in calculus may be characterized in terms of simple DEs and initial conditions. This is, in fact, one of the main reasons for their importance, and explains why such functions are central to so many applications of calculus.

EXERCISES ON 10.3

1. The functions $y_1(x)$ and $y_2(x)$ satisfy both the DE $dy/dx = Q(x, y)$ $(0 \leq x < l)$, as well as the initial condition $y(0) = y_0$. Use the DE satisfied by y_1 and y_2 to write down an expression for

$$\frac{d}{dx}(y_1 - y_2)^2$$

Assuming that the function Q satisfies an estimate of the form

$$|Q(x, y_1) - Q(x, y_2)| \leq C|y_1 - y_2|$$

where C is a positive constant, deduce that

$$\frac{d}{dx}(y_1 - y_2)^2 \le 2C(y_1 - y_2)^2$$

for x in the interval $[0, l)$. Hence use the product rule for differentiation to show that if a function σ is defined for $x \in [0, l)$ by

$$\sigma(x) = (\exp(-2Cx))(y_1 - y_2)^2$$

then $d\sigma/dx \le 0$.

Verify that $\sigma(x)$ satisfies the conditions
(a) $\sigma(0) = 0$;
(b) $\sigma(x)$ is non-increasing for $x \in [0, l)$;
(c) $\sigma(x) \ge 0$ for $x \in [0, l)$;
and deduce that in fact $\sigma(x) = 0$ for $x \in [0, l)$.

Hence show that the two solutions $y_1(x)$ and $y_2(x)$ are identical on the interval $[0, l)$. [This is a proof of uniqueness of solution, subject to a given initial condition.]

2. Complete the solution of Example 7 to show that the solution curve is given in the interval $(-2\alpha - \beta, -\alpha)$ by the equation $x = -2\alpha - \int_0^y (1/\sqrt{1 + \frac{2}{3}t^3})\, dt$. Explain why you think the improper integral $\int_0^\infty (1/\sqrt{1 + \frac{2}{3}t^3})\, dt$ is convergent.

3. Show how the single second order DE $d^2f/dx^2 = -f$ may be written as two first order equations $dy_1/dx = y_2$, $dy_2/dx = -y_1$, where $y_1 = f$. Given that f also satisfies the initial conditions $f = 0$, $df/dx = 1$ at $x = 0$, obtain the integral equations

$$y_1(x) = \int_0^x y_2(t)\, dt, \qquad y_2(x) = 1 - \int_0^x y_1(t)\, dt$$

(These integral equations were used in problem 1 of Exercises on 9.2 to derive the Taylor series for the sine and cosine functions.)

4. The function f satisfies the differential equation $d^4f/dx^4 = f$, subject to the four initial conditions $f = 1$, $df/dx = 1$, $d^2f/dx^2 = -1$, $d^3f/dx^3 = -1$ at $x = 0$. Evaluate the higher order derivatives of f, and hence write down a Taylor expansion for f about $x = 0$.

By considering the even and odd powers separately for this series expansion, identify the function $f(x)$.

5. A function $h(t)$ satisfies the DE $d^3h/dt^3 = t$, subject to the initial conditions $h = 1$, $dh/dt = 2$, $d^2h/dt^2 = 3$ at $t = 1$. Obtain the Taylor series for $h(t)$ in powers of $(t - 1)$.

10.4 Applications of differential equations

In order to fix our ideas on this subject, I would like to propose a **thought experiment**. Before taking part in this thought experiment, make sure you are familiar with Section 10.3 which dealt with the question of existence and uniqueness, and how many solutions you can expect to find, and try putting some of these ideas into practice yourself by doing a few of the problems. Then I would like you to set yourself the following task.

Imagine yourself in the position of an observer confronting a simple dynamical system for the first time. To do this, you do not need to have any prior knowledge of dynamical systems, or even any clearly defined idea of what dynamical systems **are**. Actually, it is better to approach this with a fresh mind and with the fewest possible preconceptions. I am asking you in fact to place yourself, mentally, in the place of someone seeking to set up a mathematical model of a physical process, based on observation, and using some of the mathematics we have already developed. Often the historical sequence is in the opposite direction, mathematical developments taking place after and **as a result** of new observational discoveries, rather than preceding them.

To be more specific, suppose you are observing the dynamical system known as a simple pendulum. The simple pendulum consists of a weight or bob B at the end of a string AB of length *l*. The end A of the string is attached to a fixed point, and the bob is allowed to swing freely. As in all mathematical modelling of phenomena in the real world, simplifications have to be made, so let us suppose that the string AB stays completely taut and is of fixed length *l* (no stretching) and that the pendulum swings to and fro in a **vertical** plane (Fig 10.4), rather than exhibiting circular or other kinds of motion as in Fig 10.5.

The task can be carried out as a thought experiment, though there are also some aspects that would only emerge from a real experimental investigation.

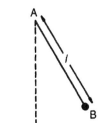

Fig 10.4 Simple pendulum.

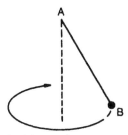

Fig 10.5 Conical pendulum.

Task

Through observations of the system consisting of a simple pendulum, create a mathematical description which allows prediction of the future motion of the system.

You: What a difficult task! You are asking me to undertake what scientists of the calibre of Galileo, Newton and many others achieved only with very great effort over an immense span of time!

Me: I take your point. But Galileo and Newton did not have our own ready access to differential equations, and other benefits of 20th century society. (Newton was, in fact, an early pioneer in the development of DEs, but in the *Principia* and most of his other writings he was able to achieve marvellous things without them.) Besides, you do not have to worry because I am going to give you a little help. We will do this together.

To start with, before we even get on to **prediction**, we need to know how to **describe** the motion of the pendulum. About the most convenient way to do this is in terms of the angle between the string of the pendulum and the vertical, that is the angle θ between AB and the dotted line in Fig 10.4. As the pendulum swings to and fro, this angle will change with time, so we shall have $\theta = \theta(t)$; θ is a function of t, and this is the first important step in our mathematical description. If we knew $\theta(t)$ in terms of t for all $t > 0$, then we would know all there is to know about the future motion of the system. The specification of the function $\theta(t)$ constitutes a **complete** description of the motion.

Of course we do not just have **one** function $\theta(t)$, we have many possible functions $\theta(t)$, depending on how we start the system off. Suppose, for example, we lift the bob (keeping AB taut) until θ has the value $\pi/4$, and then release it; then the subsequent motion will define one such function $\theta(t)$. It is important to realize here that if I carry out such an experiment, with $\theta = \pi/4$ and the bob at rest to start with, the function $\theta(t)$ will then be completely determined and may in principle be **measured** for any later value of t. If I come back in 10 seconds, or half an hour, or 2 years, and carry out the **same** experiment again, releasing the pendulum from rest with $\theta = \pi/4$ initially, then again I shall obtain just the same motion, and hence the same function $\theta(t)$. [In saying this, I am assuming that we always measure time from the instant at which the bob is released, so that $\theta(0) = \pi/4$. If we do not choose to make this simplifying assumption and, say, release at time $t = 1$ rather than $t = 0$, this will simply change our function from $\theta(t)$ to $\theta(t - 1)$.]

Another point to realize, which may in a sense be trivial but has an important bearing on the possibility of using mathematics to describe physical phenomena at all, is that if I get my friend Harry to carry out these experiments then he will obtain the same results as I do, at least to within the accuracy of his (and my) observations.

Mathematically, just as the pendulum starting at angle $\pi/4$ means that $\theta = \pi/4$ initially, so the pendulum starting from rest means that $d\theta/dt = 0$ initially. We can summarize our conclusions (which are the results of observation and there-fore experimental conclusions, albeit the results of thought experiments) by saying that if we are given that $\theta = \pi/4$ and $d\theta/dt = 0$ initially, then in principle **all** the subsequent behaviour of the system, including the function $\theta(t)$, is completely determined; and the same conclusions hold **whatever** the initial values of θ and $d\theta/dt$ happen to be. If you tell me that, for example, $\theta = \pi/3$ and $d\theta/dt = \omega$ initially (ω having some prescribed value), then I can tell you (in principle, from the results of my observations) the value of $\theta(t)$ at all subsequent times. I do not need **any** further information (such as, for example, the initial value of $d^2\theta/dt^2$) in order to determine $\theta(t)$. The initial values of θ and $d\theta/dt$ are enough, after which nature takes its course (in the form of the laws of physics) and everything else follows. The specification of an initial value of θ corresponds to starting the system off at a given angle, and the specification of an initial value of $d\theta/dt$ corresponds to starting the system off with a given initial velocity (this velocity of the bob being $l d\theta/dt$); and it is a fundamental observation that **whenever** we start the system off from a given position (or angle) and velocity, the system **always does the same thing**.

The harder part of our thinking is already done. As we watch the pendulum swing to and fro, we may do so in the confidence that a knowledge of the values of θ and $d\theta/dt$ **at this very moment** t will, by our previous observations, be enough to determine the entire subsequent history of the function θ, that is its history for all time $\geq t$. In particular, by evaluating (in principle) the second derivative, this information allows us to determine the value $d^2\theta/dt^2$ **at this very moment** t; so whenever (that is, at whatever time) we know the values of θ and $d\theta/dt$, the value of $d^2\theta/dt^2$ is uniquely determined, and the same $d^2\theta/dt^2$ results, for a given pair of values θ, $d\theta/dt$, at whatever time t we carry out this determination. We can say, with confidence founded on the results of observations, that $d^2\theta/dt^2$ is a **function** of θ and $d\theta/dt$, and we can call this function Q and express the relationship in the familiar way $d^2\theta/dt^2 = Q(\theta, d\theta/dt)$.

Not only is the proper mathematical description of our system expressed by a second order DE, but this second order DE embodies all of the predictive power for the system that we could possibly hope for. This predictive power comes from the existence and uniqueness results for second order DEs.

Let us agree to describe the **state** of the dynamical system, at a given instant of time, by the values of θ and $d\theta/dt$ at that instant of time. Then given the state at time t, the solution of the DE subject to initial conditions specifying θ and $d\theta/dt$ at time t will determine the state of the system at any subsequent time (and also at any **previous** time, if the system has been moving freely in the past). The entire dynamics is then described by the differential equation subject to initial conditions.

What can be said of the function Q on the right-hand side of the DE? In principle, Q can be determined from our experimental observations. In practice, Q is usually arrived at as a result of mathematical modelling in the light of experience and observational evidence. Simplifying assumptions as well as approximations have to be made, and there is no final definitive DE which describes the system precisely. Should we take into account the rotation of the earth? There is a pendulum suspended from the roof of the Science Museum in London for which the earth's rotation has an important effect! Should we make allowances for vibrations of the building in which the pendulum is housed, or condensation of water on the pendulum, or gravitational waves from distant parts of the universe? Even these effects might become important in certain circumstances, though obviously it is usually safe to ignore them.

In the absence of friction, air resistance, and a host of other influences that might perturb our apparatus, the basic principles of classical dynamics lead to the differential equation

$$\frac{d^2\theta}{dt^2} = -\frac{g}{l}\sin\theta$$

where l is the length (the distance AB in Fig 10.4) and g is a constant (the acceleration due to gravity, given approximately by 9.81 m s^{-2}). You can have a look at some solutions of this equation in problem 1 of Exercises on 10.4.

You may already have seen the equation written down in the approximation which holds for small values of θ, when the pendulum swings with the string close to the vertical. In that case, the equation becomes

$$\frac{d^2\theta}{dt^2} = -\frac{g}{l}\theta$$

This is an example of a **linear** DE, and such equations will be discussed in Chapter 11. The solutions of this equation describe oscillations of the pendulum with period of oscillation (time between two consecutive occasions with $\theta = \theta_{max}$, where θ_{max} is the **amplitude** of the oscillation, or maximum value of θ throughout the motion) given by $T = 2\pi\sqrt{l/g}$. We may also wish to take into account frictional effects, in which case one possible modification of the DE is

$$\frac{d^2\theta}{dt^2} = -\frac{g}{l}\theta - k\theta$$

where k is a positive constant. Further refinements and modifications are possible.

The preceding discussion is a case study of just one typical, but simple, example of a dynamical system. Similar arguments can be applied quite generally, and show that just about any dynamical system can be described in similar terms. The equations of classical dynamics are second order DEs, though for more complex systems (systems having several 'degrees of freedom') we will need more than just one function to describe the motion, and there will be a second order DE for each such function. If the system is described by functions $q_1(t), q_2(t), ...$, then the state of the system at any time t will involve a specification of $q_1, q_2, ..., dq_1/dt,$ $dq_2/dt, ...$, and by solving these DEs (which in general will be simultaneous, or 'coupled' DEs for the various functions $q_1, q_2, ...$) the state at any subsequent time can then be determined.

Not even the most ardent materialist could ever hope to describe the whole world completely in terms of classical dynamical systems and second order DEs. There are countless phenomena which do not yield their secrets to analysis of this kind. Nevertheless, there is hardly an area of scientific development where DEs have not penetrated to some degree, and mathematical modelling in areas as diverse as the study of ice accretion on overhead power lines, noise control of exhaust pipe emissions, root aeration in plants, financial forecasting, and population growth, lead to DEs almost as a matter of course. Recent progress in the understanding of 'chaos' has added a new dimension to the theory and applications of dynamical systems.

DEs are important because of their predictive power. They may be of first order, second order, or higher order, they may involve one function or several functions, which may themselves depend on one variable or several variables. (DEs in more than one variable are usually what are called 'partial differential equations', or

PDEs, and involve partial derivatives. They fall outside the scope of this book, though they are a natural extension of some of the ideas which we consider here.) The predictive power of a DE enables us to extend the solution forward (or backward) in time t. Alternatively, a DE may allow us to extend a function in space, and 'predict' what is to be found at regions of space away from the location of the initial condition(s), or to extend the function in some other variable. DEs are fundamental to so many areas of enquiry, because they are at the heart of scientific methodology.

EXERCISES ON 10.4

1. In suitable units, the differential equation for the angle θ of a simple pendulum as a function of time t may be written as

$$\frac{d^2\theta}{dt^2} + \sin\theta = 0$$

Verify by differentiation that θ satisfies the first order equation

$$\left(\frac{d\theta}{dt}\right)^2 - 2\cos\theta = c$$

where c is a constant.
   (a) In the case $c = 2$, show that $d\theta/dt = \pm 2\cos\frac{1}{2}\theta$.
   (b) In the case $c = 4$, find the maximum and minimum values of $d\theta/dt$ throughout the motion (assume $d\theta/dt > 0$). Describe the type of motion which takes place in this case.
   (c) In the case $c = 1$, find the maximum and minimum values of θ throughout the motion.
      Give a rough sketch, on a single diagram with $d\theta/dt$ as the 'y' coordinate and θ as the 'x' coordinate, of the graph of $d\theta/dt$ against θ (for $d\theta/dt > 0$), in each of the cases $c = 2$, $c = 4$, $c = 1$.
2. Use books in your library to find examples of DEs with applications to (a) biology, (b) economics and (c) industry.
3. An approximation to the DE in problem 1 in the case of θ small is $d^2\theta/dt^2 + \theta = 0$. Find a first order DE relating $d\theta/dt$ to θ in this case, and show that the plots of $d\theta/dt$ against θ are arcs of circles.

Summary

A typical first order DE looks like $df/dx = Q(x,f)$, where Q is a given function of x and f. Since, for any solution $f(x)$, the gradient at any point (x,y) of the graph $y = f(x)$ satisfies the equation $dy/dx = Q(x,y)$, a first order DE tells us what gradient the graph must have as it passes through a general point (x,y) of the xy plane. If the function Q is not too badly behaved, there will be one and only one solution curve $y = f(x)$ through any given point (x_0, y_0), or at least through any point (x_0, y_0) at which the function $Q(x,y)$ is defined. This solution curve is the graph of the solution $f(x)$ of the DE, subject to the initial condition $f(x_0) = y_0$. It may happen, even if Q is a very nice function, that the solution $f(x)$ is not defined for all values of x, but only for values of x close enough to $x = x_0$; in that case, if we go too far from x_0, the solution may diverge.

We looked at some special cases of first order DEs, including separable DEs, for which the function $Q(x, y)$ factorizes into a product of a function of x and a function of y, and some other equations for which the DE could be solved 'exactly' — that is, we could arrive at an analytic form for the general solution. In most cases, even when the function Q takes on a relatively simple form, an exact solution to the DE cannot be found. We must then be content with iterative or series methods, which may allow us to approximate the solution as closely as possible, while not providing a solution in closed form. A good idea of the behaviour of solutions may often be obtained through a graphical approach.

Most applications of DEs to science and engineering give rise to DEs of second or even higher order. Problems in dynamics, for example, usually lead to second order equations, or to systems of second order DEs, as do circuit problems in electronics, and many problems in elasticity are best described by fourth order equations. For a DE of order n, the solution will be determined by n initial conditions. Hence, for example, a fourth order DE such as $d^4 f/dx^4 = f d^2 f/dx^2 + (df/dx)^2$ will have just one solution, subject to prescribed values for f, df/dx, $d^2 f/dx^2$ and $d^3 f/dx^3$ at $x = 0$. This solution may, for example, be expressed by a power series, as in exercise 6 to follow. The solution of **linear** DEs in general will be the subject of the following chapter.

FURTHER EXERCISES

1. Solve the following first order DEs, in each case subject to the initial condition $f(0) = 1$:
   (a) $\frac{df}{dx} = \tan^{-1} x$
   (b) $\frac{df}{dx} = \frac{1}{f}$
   (c) $\frac{df}{dx} = (xf)^2$.

2. The function $y(x)$ satisfies the DE $dy/dx = 2y^{3/2}$. Use the chain rule to show that $d^2 y/dx^2 = 3y^{1/2}(dy/dx)$ and hence show that $d^2 y/dx^2 = 6y^2$.

   By further differentiation, show that $d^3 y/dx^3 = 24y^{5/2}$, and use mathematical induction to verify the results, for positive integer n,

   $$\frac{d^{2n-1} y}{dx^{2n-1}} = (2n)! y^{\frac{2n+1}{2}}, \qquad \frac{d^{2n} y}{dx^{2n}} = (2n+1)! y^{n+1}$$

   (If you are not familiar with the method of mathematical induction, simply verify these two equations for $n = 1, 2, 3$.)

   It is given that $y(x)$ satisfies the DE subject to the initial condition $y(0) = 1$. Show that $d^k y/dx^k = (k+1)!$ at $x = 0$, for all $k \geq 1$, and hence write down a power series expansion for $y(x)$.

   By considering the derivative $d(y^{-1/2})/dx$, show that $y = (1-x)^{-2}$, and verify the power series expansion by using a Taylor series.

3. Write down as many solutions as you can for each of the DEs

   $$\frac{dy}{dx} + 4x = 0, \qquad \frac{dy}{dx} + 4y = 0, \qquad \frac{dy}{dx} + 4y = x$$

4. By considering $d(e^{4x}y)/dx$, show that the solution of the DE $dy/dx + 4y = p(x)$, subject to $y(0) = 1$, is

$$y(x) = e^{-4x}\left(1 + \int_0^x e^{4t}p(t)dt\right)$$

Hence solve the differential equation $dy/dx + 4y = x^3$, subject to this initial condition.

5. The function $y(x)$ satisfies the DE $dy/dx = x^2 + y^2$, subject to the initial condition $y(0) = 0$. Verify that $y(x)$ satisfies the integral equation $y(x) = (x^3/3) + \int_0^x (y(t))^2\, dt$.

Use an iterative method which enables you to estimate the value of $y(1)$, to two decimal places.

If y is given in terms of a function z by $y = -(1/z)\, dz/dx$, verify that z satisfies the second order DE $d^2z/dx^2 + x^2z = 0$.

6. The function $f(x)$ satisfies the fourth order DE

$$\frac{d^4f}{dx^4} = f\frac{d^2f}{dx^2} + \left(\frac{df}{dx}\right)^2$$

Write down expressions for the higher order derivatives of f up to and including d^7f/dx^7. Hence write down a Taylor series for $f(x)$ up to and including power 7, given that $f = 1$, $df/dx = 0$, $d^2f/dx^2 = 1$, $d^3f/dx^3 = 0$ at $x = 0$.

11 • Linear Differential Equations

In Section 5.1 we saw that differentiation may be thought of as a linear operator, in the sense that the derivative of $Af_1 + Bf_2$ is given by

$$\frac{d}{dx}(Af_1 + Bf_2) = A\frac{df_1}{dx} + B\frac{df_2}{dx}$$

for any constants A and B. The linearity of differentiation is important in many aspects of the calculus of functions, but nowhere more so than in the theory of DEs; and especially in the theory of **linear** differential equations.

What is a linear DE? To take first of all the simplest case, a linear DE of first order is an equation $df/dx = Q(x,f)$ where for each value of x the right-hand side function depends **linearly** on f. That is, $Q = af + b$, so the DE takes the form $df/dx = af + b$. Here a and b are constants as far as f is concerned, that is, they do not depend explicitly on f; they may, however, depend on x, so we should write $a = a(x)$, $b = b(x)$, and the DE becomes $df/dx = a(x)f + b(x)$. (Of course a or b or both **may** be constant functions, or even the zero function, but these are special cases.)

A linear DE of second order is an equation $d^2f/dx^2 = Q(x,f,df/dx)$ where for each value of x the right-hand side function depends linearly on both f and df/dx. The DE then looks like $d^2f/dx^2 = a(x)df/dx + b(x)f + c(x)$, where a, b and c are given functions of x. Linear DEs of order greater than 2 are defined by an obvious extension of the same idea.

A common convention is to take all terms involving f or derivatives of f to the left-hand side of the equation. DEs of first and second order are then written, respectively, as

$$\frac{df}{dx} - a(x)f = b(x)$$

and

$$\frac{d^2f}{dx^2} - a(x)\frac{df}{dx} - b(x)f = c(x)$$

A more systematic notation is needed, particularly if we are to deal with higher order equations. We shall generally denote by $p_0(x)$ the coefficient of f, $p_1(x)$ the coefficient of $\frac{df}{dx}$, $p_2(x)$ the coefficient of $\frac{d^2f}{dx^2}$, and $p_k(x)$ the coefficient of $\frac{d^kf}{dx^k}$; the right-hand side function will be denoted by $p(x)$. Then the general first order linear DE becomes

$$p_1(x)\frac{df}{dx} + p_0(x)f = p(x)$$

and the general second order linear DE is

$$p_2(x)\frac{d^2f}{dx^2} + p_1(x)\frac{df}{dx} + p_0(x)f = p(x)$$

The general DE of order n will be

$$p_n(x)\frac{d^nf}{dx^n} + p_{n-1}(x)\frac{d^{n-1}f}{dx^{n-1}} + \cdots + p_1(x)\frac{df}{dx} + p_0(x)f = p(x)$$

As usual, we shall consider solutions $f(x)$ for x on some interval $a < x < b$. All coefficient functions $p_0, p_1, \ldots, p_n$, as well as p, will be assumed continuous on this interval, with the further condition that $p_n(x) \neq 0$; the equation may then be divided throughout by $p_n(x)$, and it will often be convenient to do this.

Examples of linear DEs are

$$(1 + x^2)\frac{df}{dx} + xf = \cos x, \quad \frac{d^3f}{dx^3} - f = 0,$$

$$(1 - x^2)\frac{d^2f}{dx^2} - 2x\frac{df}{dx} + 2f = 0 \ (-1 < x < 1), e^x\frac{d^2f}{dx^2} - (\cos x)\frac{df}{dx} = x^3$$

A linear DE will be called **homogeneous** if the right-hand side function $p(x)$ is the zero function, and **inhomogeneous** otherwise, so $d^3f/dx^3 - f = 0$ is a third order linear homogeneous DE, whereas $d^3f/dx^3 - f = 2e^x$ is third order inhomogeneous. In this chapter, I shall show how the linearity of differentiation can be exploited to determine the structure of solution sets of linear DEs, and we shall devise strategies for obtaining solutions. We start with the simplest case in which the equation is of first order.

11.1 First order linear differential equations

I shall begin by looking at three examples of first order linear DEs. These equations are superficially rather similar, but there are important differences. They are

(1) $x\frac{df}{dx} + f = 1$;
(2) $x\frac{df}{dx} - f = 1$;
(3) $\frac{df}{dx} + xf = 1$.

Each of these equations is inhomogeneous; the corresponding homogeneous DE would have zero rather than one on the right-hand side. The point about looking first of all at these three equations is that an understanding of these cases will help us to develop a strategy for coping with the general first order DE $p_1(x) \, df/dx + p_0(x)f = p(x)$.

Look now at the first of the equations $x \, df/dx + f = 1$. We have to consider this DE on an interval of x values which does not include $x = 0$. This is because we have stipulated that the coefficient of the highest power of x in a DE must never be zero; it must always be possible to divide through by this coefficient and reduce the coefficient to one.

For this reason, let us take positive values of x, that is restrict x to the interval $(0, \infty)$. Probably, with a little trial and error, you will be able to spot that the DE has a solution $f(x) = 1$. This constant function is a solution, because $x(d/dx)1 + 1 = 1$, as required, for all $x > 0$. Of course $f(x) = 1$ can hardly be the **general** solution, because there should be a whole family of solutions, corresponding to different initial conditions. (Faced with an initial condition $f(1) = 1$, or more generally $f(x_0) = 1$ for some $x_0 > 0$, the constant function $f(x) = 1$ would be **the** solution to the problem, and our task would then be over.)

To obtain the general solution, we need to develop more systematic methods than just trying out particular solutions which may or may not work; and in the first example of the three, we **can** do a lot better, by observing that the left-hand side of the DE is just $d(xf)/dx$, by the product rule for differentiation.

So, $x\frac{df}{dx} + f = 1$ can be written as $\frac{d}{dx}(xf) = 1$.

This simple observation is the key to a complete solution of the DE. Since the derivative of xf is the constant function 1, xf must be an antiderivative of 1, and it follows that

$$xf = \int 1\,dx = x + c$$

where c is a constant. Hence, on dividing throughout by x,

$$f(x) = 1 + \frac{c}{x} \quad (x > 0)$$

is the general solution. Notice how the general solution is telling us to steer clear of the point $x = 0$. Notice too that the solution we spotted earlier, $f(x) = 1$ corresponds in the solution to the special case $c = 0$.

We can, if we wish, go further and obtain the solution subject to any given initial condition $f(x_0) = y_0$; with, of course, $x_0 > 0$ to make sure that we stay within the prescribed interval $0 < x < \infty$. As an example, subject to the initial condition $f(1) = 2$, we should have $f(1) = 1 + c = 2$, leading to $c = 1$ and the particular solution

$$f(x) = 1 + \frac{1}{x} \quad (x > 0)$$

The second DE of the three, $x\,df/dx - f = 1$, is not quite so straightforward. Here you may be able to spot the single solution $f(x) = -1$, but there is no obvious way of writing the left-hand side of the equation as a derivative. However, there is a way of proceeding, even in this case. If $x\,df/dx + f$ looks like the product rule for differentiation, then $x\,df/dx - f$ looks (something) like the quotient rule. In fact

$$\frac{d}{dx}\left(\frac{f}{x}\right) = \frac{x\frac{df}{dx} - f}{x^2}$$

which agrees with the left-hand side of the DE $x\,df/dx - f = 1$, except for the factor $1/x^2$. The absence of the $1/x^2$ factor can be remedied immediately. Simply multiply both sides of the equation by $1/x^2$, and we have

$$\frac{x\frac{df}{dx} - f}{x^2} = \frac{1}{x^2} \qquad \text{or} \qquad \frac{d}{dx}\left(\frac{f}{x}\right) = \frac{1}{x^2}$$

By integration, we now have $f/x = \int (1/x^2)\,dx = -(1/x) + c$, leading to the general solution $f(x) = -1 + cx$ $(x > 0)$. Again, note how the particular solution $f(x) = -1$ appears as a special case $(c = 0)$, as it must if we really have the general solution.

What about the third DE, $df/dx + xf = 1$? Here there is no need to rule out $x = 0$, and we can look for solutions defined on the entire real line $\mathbb{R}$. In this case, it is not easy to spot any solution of the DE. To make matters worse, it is unclear how to write the left-hand side in a more amenable form, as the derivative of something. Nevertheless, let us try! We were successful in example (2) by multiplying the DE throughout by $1/x^2$. Here we can try the same thing again, not with $1/x^2$, which definitely would not work this time, but perhaps with something else, say a function $P(x)$. We leave open for the moment what function of x $P(x)$ should be. Multiplying the DE throughout by $P(x)$ gives

$$P(x)\frac{df}{dx} + xP(x)f = P(x)$$

We **hope** the left-hand side will be the derivative of something. The derivative of what? Well, guided by example (2), the most obvious answer to this would be the derivative of the product $P(x)f$. Since, by the product rule

$$\frac{d}{dx}(P(x)f) = P(x)\frac{df}{dx} + \frac{dP(x)}{dx}f$$

we have got the first term right on the left-hand side, and the second term will also be right provided it is true that

$$\frac{d}{dx}P(x) = xP(x)$$

Here we are getting somewhere! The equation $dP/dx = xP$ is a DE for the function P. It is also a first order linear homogeneous DE (homogeneous because $dP/dx - xP = 0$). It is even a **separable** DE, of the kind that we have already learnt how to solve in Section 10.1. Assuming $P > 0$, our standard method for solution of equations of this kind $((1/y)\,dy/dx = x$; consider the derivative with respect to x of $\ln(y)$, where $\ln(y)$ is an antiderivative of $1/y)$ gives

$$\frac{d}{dx}(\ln P) = \frac{1}{P}\frac{dP}{dx} = \frac{1}{P} \cdot xP = x$$

Hence a possible choice of the function P is given by

$$\ln(P) = \int x\,dx \quad \text{or} \quad P(x) = \exp(x^2/2)$$

(There is no need to find the most general function $P(x)$, which would be a constant multiple of this; one such function is sufficient.)

Now we know what function $P(x)$ to take, we can multiply example (3), $df/dx + xf = 1$, throughout by this function to obtain

$$e^{\frac{x^2}{2}}\frac{df}{dx} + xe^{\frac{x^2}{2}}f = e^{\frac{x^2}{2}}$$

Having written the DE in this form, the left-hand side **must** be the derivative of something. That is what $P(x)$ was designed to do, and that is why we made this particular choice of $P(x)$. We also know what the left-hand side is the derivative **of**. $P(x)$ was chosen to make the left-hand side the derivative of $P(x)f$; with this choice of P, the left-hand side can **only** be the derivative of $e^{\frac{x^2}{2}}f$. Even so, it is a useful check on our calculations to verify, using the product rule, that the DE can now be written as

$$\frac{d}{dx}(e^{\frac{x^2}{2}}f) = e^{\frac{x^2}{2}}$$

Since there is no obvious antiderivative of $e^{\frac{x^2}{2}}$, we can use the fundamental theorem of calculus to make one: $\int_0^x e^{\frac{t^2}{2}} dt$. (If an initial condition is imposed at $x = x_0$, a better choice might be $\int_{x_0}^x e^{\frac{t^2}{2}} dt$.)

Integrating the equation, we now have

$$e^{\frac{x^2}{2}}f = \int_0^x e^{\frac{t^2}{2}} dt + c$$

where c is a constant. The general solution of the DE is given by

$$f(x) = e^{-\frac{x^2}{2}} \int_0^x e^{\frac{t^2}{2}} dt + ce^{-\frac{x^2}{2}}$$

No wonder you were (probably) unsuccessful at spotting a particular solution! There are many particular solutions, obtained by giving different values to the constant c, but none of them are easy to find by guesswork. Note that $f(0) = c$, so that the constant c can be determined by an initial condition at $x = 0$.

The basic idea which we used in solving (2) and (3) was to multiply the DE by an appropriate factor in order to convert the left-hand side to the derivative of something. This idea was behind the solution of (1) as well, except that there the left-hand side was already in a suitable form, so that no multiplying factor was necessary.

The same idea can be applied quite generally to linear first order DEs. The function $P(x)$ by which the DE is multiplied in order to convert it into a suitable derivative form is called an **integrating factor** for the DE. In problem 1 of Exercises on 11.1, you are asked to verify the integrating factor in the general case. To save you the trouble of going through the same kind of argument each time you need to solve a first order linear DE, I have summarized the result in the following general strategy.

SOLUTION
Write the DE in the form $df/dx + p_0(x)f = p(x)$. (This involves dividing the equation throughout by the coefficient of df/dx, if this is not already 1.)

If $P_0(x)$ is an antiderivative of $p_0(x)$, an integrating factor of the DE is now $P(x) = \exp P_0(x)$, i.e. the integrating factor is given by $P(x) = \exp \int p_0(x) dx$. The integration of $p_0(x)$ should be carried out explicitly if possible; if this cannot be done, then leave $P(x)$ in a general form such as $P(x) = \exp \int_{x_0}^x p_0(t) dt$. Any possible simplifications (e.g. involving identities for exponential and logarithmic functions) to the expression for $P(x)$ should be carried out before proceeding.

Now multiply the DE throughout by the integrating factor $P(x)$. The DE will then be expressible in the form

$$\frac{d}{dx}(P(x)f) = p(x)P(x)$$

This equation may then be integrated to give

$$P(x)f = \int p(x)P(x)\,dx + c$$

where c is a constant.

Carrying out the integration, or writing an expression for the integral such as $\int_{x_0}^{x} p(t)P(t)\,dt$, the general solution is finally obtained on dividing through by $P(x)$; the constant c will then be determined by an initial condition, if one is given.

Rather than trying to remember any general formula for the solution, it is best just to bear in mind the expression for the integrating factor (exponential of the integral of the coefficient of f **after** the coefficient of df/dx has been reduced to one) as well as the general idea behind the use of integrating factors. The following examples illustrate the method in action.

⊛ *Example I*

We solve the DE $x\,df/dx + 3f = 5x^2$ $(x > 0)$, subject to the initial condition $f(1) = 2$.

Reducing to one the coefficient of df/dx, we must divide through by x, giving

$$\frac{df}{dx} + \frac{3}{x}f = 5x$$

The coefficient of f is now $3/x$, of which an antiderivative is $3\ln(x)$. Hence an integrating factor for the DE is

$$P(x) = \exp \int \frac{3}{x}\,dx = \exp(3\ln(x))$$

Although this is a correct integrating factor, it should be simplified using properties of the exponential and logarithm. Thus

$$P(x) = \exp(\ln(x^3)) = x^3$$

Multiplying through by this factor, the DE becomes

$$x^3\frac{df}{dx} + 3x^2 f = 5x^4$$

or

$$\frac{d}{dx}(x^3 f) = 5x^4$$

Hence $x^3 f = \int 5x^4\,dx = x^5 + c$, and we obtain the general solution $f(x) = x^2 + c/x^3$.

The initial condition $f(1) = 2$ gives $1 + c = 2$, so that $c = 1$ and our solution to the problem is $f(x) = x^2 + 1/x^3$.

● *Example 2*

Consider now the first order linear DE $(1 + x^2)\, df/dx - 2xf = 1$. We look for the general solution.

First write the equation in the form

$$\frac{df}{dx} - \frac{2x}{(1 + x^2)} f = \frac{1}{(1 + x^2)}$$

The coefficient of f is $-2x/(1 + x^2)$; hence there is an integrating factor:

$$\exp \int -\frac{2x}{(1 + x^2)}\, dx = \exp(-\ln(1 + x^2)) = \exp \ln(1/(1 + x^2)) = 1/(1 + x^2)$$

Notice that the minus sign in the coefficient of f was crucial here. Had we taken instead $\exp(+ \int [2x/(1 + x^2)]\, dx)$, we should have obtained not $1/(1 + x^2)$ but $(1 + x^2)$, which is completely incorrect as an integrating factor.

Now multiplying the DE by the correct integrating factor $1/(1 + x^2)$ gives

$$\frac{1}{(1 + x^2)}\frac{df}{dx} - \frac{2x}{(1 + x^2)^2} f = \frac{1}{(1 + x^2)^2}$$

or

$$\frac{d}{dx}\left(\frac{1}{(1 + x^2)} f\right) = \frac{1}{(1 + x^2)^2}$$

It is not altogether obvious how to find an antiderivative of the right-hand side. If you have a good memory, you may recall Example 6 of Section 8.3, where we found that

$$\int \frac{x^2}{(1 + x^2)^2}\, dx = -\frac{x}{2(1 + x^2)} + \frac{1}{2}\tan^{-1} x + \text{const}$$

The required integral is closely related to this, since we can write

$$\frac{1}{(1 + x^2)^2} = \frac{(1 + x^2) - x^2}{(1 + x^2)^2} = \frac{1}{(1 + x^2)} - \frac{x^2}{(1 + x^2)^2}$$

Hence on integrating the DE, we have

$$\frac{1}{(1 + x^2)} f = \int \frac{1}{(1 + x^2)^2}\, dx + c = \int \frac{1}{(1 + x^2)}\, dx - \int \frac{x^2}{(1 + x^2)^2}\, dx + \text{const}$$

$$= \tan^{-1} x - \left(-\frac{x}{2(1 + x^2)} + \frac{1}{2}\tan^{-1} x\right) + \text{const}$$

$$= \frac{1}{2}\left(\frac{x}{(1 + x^2)} + \tan^{-1} x\right) + c$$

from which the solution

$$f(x) = \frac{1}{2}(x + (1 + x^2)\tan^{-1} x) + c(1 + x^2)$$

follows. Do not despair if you are unable to carry out some of the integration explicitly. Two possible remedies of this situation are:

(a) consult one of the many reference books in your library which contains tables of integrals; or

(b) use the fundamental theorem of calculus to write an antiderivative as a definite integral, in this case leading to the solution

$$f(x) = (1 + x^2) \int_0^x \frac{1}{(1 + t^2)^2} dt + c(1 + x^2)$$

In following through the examples of first order linear DEs that we have looked at so far in this chapter, you will probably have noticed that in each case the general solution takes the form of a sum of a function f_0 and a constant multiple of another function f_1. That is, there is a common pattern that the general solution looks like $f(x) = f_0(x) + cf_1(x)$, where f_0 and f_1 are two functions and c is a constant. It is not difficult to verify, using the method of integrating factor, that the general solution of a first order linear DE must **always** have this form.

The function f_0, being the solution corresponding to $c = 0$, is a particular solution of the DE. What can be said about the function f_1? Well, if $f_0 + cf_1$ is the general solution of the DE $p_1 \, df/dx + p_0 f = p$, where for simplicity of notation I have suppressed the dependence on x of the coefficient functions p_0, p_1 and the right-hand side function p, we can substitute $c = 1$, for example, in which case

$$p_1 \frac{d}{dx}(f_0 + f_1) + p_0(f_0 + f_1) = p$$

Separating out the terms involving f_0 from those involving f_1, this gives

$$\left\{ p_1 \frac{df_0}{dx} + p_0 f_0 \right\} + p_1 \frac{df_1}{dx} + p_0 f_1 = p$$

However, we know that f_0 is a solution of the DE, so that the terms within the braces together equate to the function p. Cancelling with the right-hand side of the equation, we can conclude that

$$p_1 \frac{df_1}{dx} + p_0 f_1 = 0$$

That is, f_1 satisfies the **homogeneous** DE $p_1 \, df/dx + p_0 f = 0$, which is obtained from the original DE by replacing the right-hand side by the zero function. So the general solution is always of the form $f(x) = f_0(x) + cf_1(x)$, where f_0 is the particular solution of the **inhomogeneous** DE corresponding to $c = 0$, and f_1 is a solution of the **homogeneous** DE.

In fact, in writing down the general solution, **any** particular solution of the inhomogeneous DE will do, and **any** solution (except the so-called trivial solution $f_1(x) = 0$) of the homogeneous DE will do for f_1. For if f_0 and f_1 are such solutions, it follows that

$$p_1 \frac{d}{dx}(f_0 + cf_1) + p_0(f_0 + cf_1) = \left\{ p_1 \frac{df_0}{dx} + p_0 f_0 \right\} + c \left\{ p_1 \frac{df_1}{dx} + p_0 f_1 \right\}$$

$$= p + c.0 = p$$

so that $f_0 + cf_1$ satisfies the inhomogeneous DE; and $f_0 + cf_1$ must be the **general** solution, since the constant c can be adjusted to match any given initial condition, so that every solution of the DE is included.

In Section 11.2, we shall look at the structure of solution sets for linear DEs of arbitrary order, and see how results in the first order case fit into this more general framework. As the following example shows, an understanding of the nature of the general solution may also lead to a short cut in solving the DE.

▨ *Example 3*

The first order linear DE $df/dx - f = x$ may be solved by use of an integrating factor. An alternative approach, based on the fact that $f(x) = f_0(x) + cf_1(x)$, where f_0 is a particular solution of the DE and f_1 satisfies the homogeneous equation, is as follows.

For f_0, in view of the fact that the right-hand side is a first degree polynomial, it seems reasonable to look for a solution of the form $f_0(x) = ax + b$, where a and b are constants to be determined. Substituting into the DE gives

$$\frac{d}{dx}(ax + b) - (ax + b) = x$$

which leads to

$$-ax + (a - b) = x$$

Regarding the right-hand side as $1.x + 0$, and equating coefficients, $f_0(x) = ax + b$ will satisfy the DE provided $-a = 1$ and $a - b = 0$. That is, $a = b = -1$, and we have the particular solution

$$f_0(x) = -(x + 1)$$

For a (non-trivial) solution of the homogeneous DE $df/dx - f = 0$, we take $f_1(x) = e^x$. Hence the general solution of $df/dx - f = x$ is given by

$$f(x) = -(x + 1) + c e^x$$

The same result can be obtained by use of an integrating factor. It should be noted at this point that two different methods may quite possibly lead to general solutions which appear superficially different. For example, another expression for the general solution in this example is $f(x) = (e^x - 1 - x) + c e^x$. There is some advantage in writing the solution in this form, since this represents the solution of the DE subject to the initial condition $f(0) = c$.

In comparing two different expressions for the general solution, it should be remembered that it is a question of comparing not two solutions but two solution sets. The set of solutions

$$\{-(x + 1) + c e^x; \quad c \in \mathbb{R}\}$$

is in fact identical to the set of solutions

$$\{e^x - 1 - x + c e^x; \quad c \in \mathbb{R}\}$$

and both describe the same general solution. This becomes clear on observing that a relabelling of the constant, $c \rightarrow c + 1$, sends one expression into the other.

1. Verify that the function $P(x) = \exp \int p_0(x)\,\mathrm{d}x$ is an integrating factor for the DE $\mathrm{d}f/\mathrm{d}x + p_0(x)f = p(x)$, in the sense that the DE, on multiplying throughout by $P(x)$, may be written as

$$\frac{\mathrm{d}}{\mathrm{d}x}(P(x)f) = p(x)P(x)$$

Use this result to find the general solution of the DE $\mathrm{d}^2f/\mathrm{d}x^2 + 2xf = x$. Demonstrate an alternative method of solving the same DE.

2. The function $y(x)$ satisfies the first order linear DE $\mathrm{d}y/\mathrm{d}x + p_0(x)y = p(x)$, where the functions p_0 and p are continuous and bounded on a given interval. Writing the DE in the form $\mathrm{d}y/\mathrm{d}x = Q(x,y)$, verify that the function Q satisfies an estimate

$$|Q(x,y_1) - Q(x,y_2)| \leq \mathrm{const}|y_1 - y_2|$$

Use this result to confirm the existence and uniqueness results of Section 10.3, that there exists one and only one solution of the DE on the given interval, subject to an initial condition $y(x_0) = y_0$.

[The standard existence/uniqueness theorem of Section 10.3 for first order DEs only guarantees the existence of a unique solution, subject to an initial condition $f(x_0) = y_0$, in some neighbourhood of the initial point $x = x_0$. Away from that point, the solution may well diverge at some value of x. For **linear** DEs, $\mathrm{d}f/\mathrm{d}x + p_0 f = p$, the situation is different (and better!); solutions remain continuous and bounded in any finite interval, so long as the coefficient functions are continuous and bounded.]

3. Let f_0 be the solution of $\mathrm{d}f/\mathrm{d}x + p_0(x)f = p(x)$, subject to $f_0(x_0) = 0$, and let f_1 be the solution of $\mathrm{d}f/\mathrm{d}x + p_0(x)f = 0$, subject to $f_1(x_0) = 1$. Verify that $f(x) = f_0(x) + y_0 f_1(x)$ is the solution of $\mathrm{d}f/\mathrm{d}x + p_0(x)f = p(x)$, subject to the condition $f(x_0) = y_0$.

Taking $x_0 = 0$, determine the functions f_0 and f_1 for the DE

$$\frac{\mathrm{d}f}{\mathrm{d}x} - 2f = 2x(1 - x)$$

and hence solve the DE subject to the initial condition $f(0) = 2$.

11.2 Linear differential equations — general theory

The kind of arguments that helped to clarify the structure of solution sets in the case of first order linear DEs can also be applied to linear DEs in general. These arguments are heavily dependent on the linearity of differentiation.

As in the first order case, let us suppose that we can find one particular solution of the DE. I shall call this solution $f_0(x)$. It will not usually be so easy to find particular solutions as for first order equations — there are quite simple-looking second order DEs, for example, which have no solutions expressible in terms of the known standard functions or their integrals. Still, we have to start somewhere, so let us suppose that somehow or other we have managed to find just one solution f_0 of the DE. (Later we shall look at some strategies for finding such a solution f_0.)

To save complications of notation, I shall consider the second order case, though the argument is quite general. Suppose, then, that $f_0(x)$ is a particular solution of

$$\frac{d^2 f}{dx^2} + p_1 \frac{df}{dx} + p_0 f = p$$

Then

$$\frac{d^2 f_0}{dx^2} + p_1 \frac{df_0}{dx} + p_0 f_0 = p$$

(Here p_0, p_1 and p are all functions of x, but again for convenience of notation I am not indicating this dependence on x explicitly. I have also simplified the equation by dividing through by the coefficient of $d^2 f/dx^2$.)

Subtracting the two equations and collecting similar terms together, we now have

$$\left(\frac{d^2 f}{dx^2} - \frac{d^2 f_0}{dx^2}\right) + p_1 \left(\frac{df}{dx} - \frac{df_0}{dx}\right) + p_0(f - f_0) = p - p = 0$$

which becomes, on using the linear property of differentiation,

$$\frac{d^2}{dx^2}(f - f_0) + p_1 \frac{d}{dx}(f - f_0) + p_0(f - f_0) = 0$$

Hence $f - f_0$ satisfies the **homogeneous** DE, obtained by replacing the right-hand side function by zero in the original DE.

What we have shown is that if f is any solution of the DE then the difference $f - f_0$, where f_0 is our particular solution, will satisfy the homogeneous DE. Another way of putting this result is that any solution of the DE is the sum of f_0, the particular solution, and a solution of the homogeneous DE. Since f_0 is regarded as known, it remains only to find the general solution of the homogeneous DE (Fig 11.1). This argument applies just as well to linear DEs of any order, so that the schematic equation works quite generally.

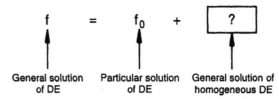

Fig 11.1   General solution of a linear DE.

We proceed, now, to the solution of the homogeneous DE, which takes the form in the second order case:

$$\frac{d^2 f}{dx^2} + p_1 \frac{df}{dx} + p_0 f = 0$$

Again, the linearity of differentiation plays a crucial rule. The key fact about linear **homogeneous** DEs (whether of second order or not) is that, if f_1 and f_2 are two solutions, then $Af_1 + Bf_2$ is also a solution, for any constants A and B.

To verify this for second order equations, suppose we have two solutions f_1, f_2, of the homogeneous DE. Then

$$\frac{d^2f_1}{dx^2} + p_1\frac{df_1}{dx} + p_0f_1 = 0$$

and

$$\frac{d^2f_2}{dx^2} + p_1\frac{df_2}{dx} + p_0f_2 = 0$$

Multiplying the first equation by A, the second by B, and adding, we can collect similar terms together to give

$$\left(A\frac{d^2f_1}{dx^2} + B\frac{d^2f_2}{dx^2}\right) + p_1\left(A\frac{df_1}{dx} + B\frac{df_2}{dx}\right) + p_0(Af_1 + Bf_2) = 0$$

Taking account of the linearity of differentiation, this becomes

$$\frac{d^2}{dx^2}(Af_1 + Bf_2) + p_1\frac{d}{dx}(Af_1 + Bf_2) + p_0(Af_1 + Bf_2) = 0$$

Hence $Af_1 + Bf_2$ is also a solution of the homogeneous DE, as stated.

This is really all we need in order to write down a general expression for the solution of the homogeneous DE. The following example shows in a simple case how this can be done.

✣ *Example 4*

General solution of homogeneous DE: we consider the second order linear homogeneous DE $d^2f/dx^2 - f = 0$. In order to apply the above result, we will need two solutions f_1, f_2, of this DE. Take the first solution to be $f_1(x) = e^x$, and the second to be $f_2(x) = e^{-x}$. Both functions may easily be verified to satisfy the DE.

It would be no use taking two solutions of which one was a constant multiple of the other, say e^x and $2e^x$, because in that case the combination $(Ae^x + B.2e^x = (A + 2B)e^x)$ is just a multiple of one solution, and we might just as well write Af_1 rather than $Af_1 + Bf_2$.

We know that $Ae^x + Be^{-x}$ is a solution of $d^2f/dx^2 - f = 0$, since both e^x and e^{-x} satisfy this equation. Is it true that $Ae^x + Be^{-x}$ is the **general** solution? To show that $Ae^x + Be^{-x}$ satisfies the DE is relatively straightforward: it is only necessary to substitute $f = Ae^x + Be^{-x}$ into the DE to verify this; but to show that $Ae^x + Be^{-x}$ is the general solution requires a different argument altogether — we need to show that **every** solution of $d^2f/dx^2 - f = 0$ is of the form $f(x) = Ae^x + Be^{-x}$, for some constants A and B. Suppose, then, that $f(x)$ is any function satisfying the equation $d^2f/dx^2 - f = 0$. Let us try to show that $f(x) = Ae^x + Be^{-x}$ for some constants A and B. Now, if it is really true that $f = Ae^x + Be^{-x}$, then by differentiation it must also be true that $df/dx = Ae^x - Be^{-x}$. This tells us what the constants A and B must be. Regarding the equation for f and the equation for df/dx as algebraic equations for A and B, we can use them to solve for A and B. Eliminating B by adding the two equations gives $2Ae^x = f + df/dx$, so that

$$A = \frac{1}{2}e^{-x}\left(f + \frac{df}{dx}\right)$$

and eliminating A leads to

$$B = \frac{1}{2}e^x\left(f - \frac{df}{dx}\right)$$

These two equations are the key to proving that the general solution is $Ae^x + Be^{-x}$. What we have to do is to start the proof by **defining** two functions $\frac{1}{2}e^{-x}(f + df/dx)$ and $\frac{1}{2}e^x(f - df/dx)$ and using $d^2f/dx^2 - f = 0$ to **prove** that these two functions are constant functions. It is then relatively easy to complete the proof using a little simple algebra. The argument can be set out as follows.

PROOF

Proof that if $d^2f/dx^2 - f = 0$ **then** $f(x) = Ae^x + Be^{-x}$. Let $f(x)$ be any function satisfying $d^2f/dx^2 - f = 0$. We prove that the two functions $\frac{1}{2}e^{-x}(f + df/dx)$, $\frac{1}{2}e^x(f - df/dx)$ are constant functions by showing in each case that the derivative is zero. (Remember from Section 6.2 that any function having zero derivative on an interval is a constant function on the interval.)

Using the product rule, the derivative of the first function is

$$\frac{d}{dx}\left\{\frac{1}{2}e^{-x}\left(f + \frac{df}{dx}\right)\right\} = \frac{1}{2}e^{-x}\frac{d}{dx}\left(f + \frac{df}{dx}\right) + \frac{1}{2}\left(f + \frac{df}{dx}\right)\frac{d}{dx}e^{-x}$$

$$= \frac{1}{2}e^{-x}\left(\frac{df}{dx} + \frac{d^2f}{dx^2}\right) - \frac{1}{2}e^{-x}\left(f + \frac{df}{dx}\right)$$

$$= \frac{1}{2}e^{-x}\left(\frac{d^2f}{dx^2} - f\right) = 0$$

since f is assumed to satisfy $d^2f/dx^2 - f = 0$.

The derivative of the second function is

$$\frac{d}{dx}\left\{\frac{1}{2}e^x\left(f - \frac{df}{dx}\right)\right\} = \frac{1}{2}e^x\left(\frac{df}{dx} - \frac{d^2f}{dx^2}\right) + \frac{1}{2}e^x\left(f - \frac{df}{dx}\right)$$

$$= -\frac{1}{2}e^x\left(\frac{d^2f}{dx^2} - f\right) = 0$$

Hence both functions are constant functions, and so we can write

$$\left.\begin{array}{l}\frac{1}{2}e^{-x}\left(f + \frac{df}{dx}\right) = A \\[2mm] \frac{1}{2}e^x\left(f - \frac{df}{dx}\right) = B\end{array}\right\}$$

for some constants A and B. (These are the same equations as I wrote down earlier! The difference is that now we have proved them from $d^2f/dx^2 - f = 0$, rather than from the general solution which we are trying to arrive at.)

Finally, it is a matter of simple algebra to verify from the equations for A and B that

$$Ae^x + Be^{-x} = \frac{1}{2}\left(f + \frac{df}{dx}\right) + \frac{1}{2}\left(f - \frac{df}{dx}\right) = f$$

as required. So $f(x) = Ae^x + Be^{-x}$ **is** the general solution, as we set out to show.

The same kind of proof can be used for any second order linear homogeneous DE. If f_1 and f_2 are two solutions, assumed to be **independent** in the sense that neither is a constant multiple of the other, then the general solution of the homogeneous DE is $f(x) = Af_1(x) + Bf_2(x)$. The following example carries out the proof in another important special case.

❋ *Example 5*

Consider now the second order linear homogeneous DE $d^2f/dx^2 + f = 0$. It is easy to verify that the functions $f_1(x) = \cos x$ and $f_2(x) = \sin x$ are solutions. Certainly these are two independent solutions, so that the general solution should be $f(x) = A\cos x + B\sin x$. To **prove** this, note that $f = A\cos x + B\sin x$ and $f' = -A\sin x + B\cos x$ would imply $A = f\cos x - f'\sin x$, $B = f\sin x + f'\cos x$.

Starting from the other end of the problem, let f be any function satisfying $d^2f/dx^2 + f = 0$. Then

$$\frac{d}{dx}\{f\cos x - f'\sin x\} = f'\cos x - f\sin x - f''\sin x - f'\cos x$$
$$= -\sin x(f'' + f) = 0$$

from the DE. Similarly

$$\frac{d}{dx}\{f\sin x + f'\cos x\} = \cos x(f'' + f) = 0$$

as well, so we can write $f\cos x - f'\sin x = A$, $f\sin x + f'\cos x = B$, and from simple algebra we have $A\cos x + B\sin x = f$ as required.

The general solution of second order linear DEs may also be obtained through an application of the existence/uniqueness theorem, which at the same time helps to show why two solutions are required, and not, say, one or three. We know that a second order DE requires **two** initial conditions in order to specify the solution uniquely. To simplify matters, I shall choose these initial conditions in a rather special way.

Given a linear homogeneous DE $d^2f/dx^2 + p_1\,df/dx + p_0 f = 0$, let f_1 be the solution subject to initial conditions $f_1 = 1$, $df_1/dx = 0$ at $x = x_0$, and let f_2 be the solution subject to $f_2 = 0$, $df_2/dx = 1$ at $x = x_0$. Now let f be **any** solution of the homogeneous DE, with, say, $f = c_0$ and $df/dx = c_1$ at $x = x_0$. We know that the combination $c_0 f_1 + c_1 f_2$ satisfies the DE, from our previous result, since f_1 and f_2 are solutions and the DE is homogeneous. Moreover, $c_0 f_1 + c_1 f_2$ satisfies the DE with the **same** initial conditions as f, since at $x = x_0$ we have

$$c_0 f_1 + c_1 f_2 = c_0.1 + c_1.0 = c_0$$

and at $x = x_0$ we have

$$\frac{d}{dx}(c_0 f_1 + c_1 f_2) = c_0 \frac{df_1}{dx} + c_1 \frac{df_2}{dx} = c_0.0 + c_1.1 = c_1$$

(Note again that linearity of differentiation is at the heart of the argument.) Since the solutions f and $c_0 f_1 + c_1 f_2$ satisfy the same initial conditions at $x = x_0$, they must be the same solution, $f(x) = c_0 f_1(x) + c_1 f_2(x)$.

Hence the solution $f(x)$ is given by $f = A f_1 + B f_2$, where $A = c_0$ and $B = c_1$. Since f was **any** solution of the homogeneous DE, we have shown that $A f_1 + B f_2$ is the general solution of this equation.

The same kind of argument can be applied to homogeneous linear DEs of higher order. In the case of order 3, we need three initial conditions to specify a solution. The solutions f_1, f_2, f_3 could, for example, be defined by the initial conditions, at $x = x_0, f_1 = 1, f'_1 = 0, f''_1 = 0, f_2 = 0, f'_2 = 1, f''_2 = 0; f_3 = 0, f'_3 = 0, f''_3 = 1$. This choice of initial conditions is only (possibly) the most convenient — the main point is that the three solutions f_1, f_2, f_3 should be independent in the sense that no solution is a combination of the other two. (For example, we should not want $f_1 = e^x, f_2 = e^{-x}, f_3 = \cosh x$; in this case the third solution would not be 'new' but merely a combination of the other two, $\cosh x = \frac{1}{2}e^x + \frac{1}{2}e^{-x}$. For readers familiar with basic ideas of linear algebra, the notion of 'combination' of solutions corresponds to the algebraic notion of linear combination, and 'independence' refers to linear independence.)

For a **third order** homogeneous DE, the general solution is made up of a combination of **three** independent solutions, $f = A f_1 + B f_2 + C f_3$. For a (linear) homogeneous DE of order n, the general solution is of the form $f = \sum_1^n A_k f_k$, where the coefficients A_k are all constants, and the f_k are n independent solutions of the homogeneous DE.

Remembering that any solution of a linear **inhomogeneous** DE is the sum of a particular solution f_0, and a solution of the corresponding **homogeneous** DE for which the right-hand side function is zero, we can set out the following strategy for solving linear DEs of arbitrary order.

SOLUTION

Let f_0 be a particular solution of the DE. If the DE is homogeneous take $f_0(x) = 0$. If the DE is inhomogeneous, then a particular solution can often be discovered by trial and error, using as a guide the nature of the right-hand side function $p(x)$. (For example, if $p(x)$ is a polynomial, it may be possible to find a polynomial solution for f_0.)

Now let $f_1, f_2, ..., f_n$ be n independent solutions of the homogeneous DE obtained by setting $p(x) = 0$. Then the general solution of the linear DE is given by

$$f(x) = f_0(x) + A_1 f_1(x) + A_2 f_2(x) + ... + A_n f_n(x)$$

where the coefficients $A_1, A_2, ..., A_n$ are all constants. These coefficients are uniquely determined by n initial conditions, applied at a point $x = x_0$ to the function f.

As a special case, with $A = A_1, B = A_2$, the general solution of a second order linear DE is given by

$$f(x) = f_0(x) + A f_1(x) + B f_2(x)$$

where f_0 is a particular solution of the DE (any solution will do) and f_1, f_2 are any two independent solutions of the homogeneous DE. The constants A, B, may then be determined by two initial conditions.

The main problem in implementing the strategy, in the case of a specific linear DE, will usually be that of finding the particular solution f_0, and sufficiently many solutions, n for a DE of order n, of the homogeneous DE. This task will sometimes be so difficult that it cannot be carried out. There may be no solution in closed form at all, if we are confined to the usual, standard functions of calculus. In that case, the only viable approach may be a numerical one, though in most cases an initial, theoretical attack on the problem will at least yield qualitative properties of solutions, and this may be all that is required.

Fortunately, there is an important class of linear DEs, that of **constant coefficient** equations, for which the strategy for solution can usually be carried out in practice. Constant coefficient DEs provide, moreover, an excellent introduction to the study of linear DEs in general. It is to this type of equation that we shall turn in the following section.

The following example illustrates the application of the ideas of this section to a concrete situation.

⊛ *Example 6*

Solve the second order linear DE

$$x^2 \frac{\mathrm{d}^2 f}{\mathrm{d}x^2} + x \frac{\mathrm{d}f}{\mathrm{d}x} - f = \ln(x) \quad (x > 0)$$

subject to the initial conditions $f = 0$, $\mathrm{d}f/\mathrm{d}x = 0$ at $x = 1$.

To solve this DE, we need one solution of the inhomogeneous DE and two solutions of the homogeneous DE. For the inhomogeneous DE, it would seem that a particular solution f_0 must somehow involve the logarithmic function; otherwise we could not expect to obtain $\ln(x)$ by substituting $f = f_0$ on the left-hand side. The simplest thing is to try $f_0(x) = \ln(x)$.

Now $f_0 = \ln(x)$, substituted into the DE, results in

$$x^2 \frac{\mathrm{d}^2 f_0}{\mathrm{d}x^2} + x \frac{\mathrm{d}f_0}{\mathrm{d}x} - f_0 = x^2 \cdot \left(-\frac{1}{x^2}\right) + x \cdot \left(\frac{1}{x}\right) - \ln(x) = -\ln(x)$$

This is very nearly what we want, but has the wrong sign. This is easily remedied, because the sign of the left-hand side can be altered by replacing $f_0 = \ln(x)$ by $f_0 = -\ln(x)$, so $f_0(x) = -\ln(x)$ is a particular solution of the DE.

The next step is to find two solutions of the homogeneous equation $x^2 \, \mathrm{d}^2 f/\mathrm{d}x^2 + x \, \mathrm{d}f/\mathrm{d}x - f = 0$. This is rather a matter of trial and error. However, in this example we are very much helped by the observation that the substitution for f of any **power** of x will result, on the left-hand side, in three terms each of which is a constant multiple of that power. For example, if $f = x^4$ then $\mathrm{d}f/\mathrm{d}x = 4x^3$ and $\mathrm{d}^2 f/\mathrm{d}x^2 = 12x^2$, so that $x^2 \, \mathrm{d}^2 f/\mathrm{d}x^2 + x \, \mathrm{d}f/\mathrm{d}x - f = 12x^4 + 4x^4 - x^4$. Although this is not zero, as we would need for a true solution of the homogeneous DE, it is possible that a judicious choice of that power will lead to three terms which sum to zero. Inspired by this hope, let us try $f = x^c$. Then $\mathrm{d}f/\mathrm{d}x = cx^{c-1}$,

$d^2f/dx^2 = c(c-1)x^{c-2}$, and $x^2 d^2f/dx^2 + x df/dx - f = c(c-1)x^c + cx^c - x^c = (c^2 - 1)x^c$.

For this to be zero, we need c to be either $+1$ or -1, so we have found two solutions x^c of the homogeneous DE, corresponding to the two possible values of c. Our two solutions are $f_1(x) = x$ and $f_2(x) = x^{-1}$.

Following the strategy for solution of linear DEs, we have the general solution $f(x) = -\ln(x) + Ax + Bx^{-1}$. It remains only to use the two initial conditions to determine the constants A and B.

$$\left.\begin{array}{ll} f(1) = 0 & \text{implies } -\ln(1) + A + B = 0, \text{ or } A + B = 0 \\ f'(1) = 0 & \text{implies } -1/1 + A - B/1^2 = 0, \text{ or } A - B = 1 \end{array}\right\}$$

Hence $A = \frac{1}{2}$, $B = -\frac{1}{2}$, and we have, finally, the solution $f(x) = \frac{1}{2}(x - 1/x) - \ln(x)$.

EXERCISES ON 11.2

1. Which of the following second order DEs are linear? For those DEs which are linear, identify which are homogeneous and which are inhomogeneous:

$$\frac{d^2f}{dx^2} - x^2f = \cos x, \quad \frac{d^2f}{dx^2} - xf^2 = \cos x, \quad x\frac{d^2f}{dx^2} - f = f^2,$$

$$\frac{d^2f}{dx^2} - f = x^2, \quad \frac{d^2f}{dx^2} = x^2f$$

2. Using the argument of Examples 4 and 5 as a model, show that if k is a positive constant then the general solution of $d^2f/dx^2 - k^2f = 0$ is $f = Ae^{kx} + Be^{-kx}$, and the general solution of $d^2f/dx^2 + k^2f = 0$ is $f = A\cos kx + B\sin kx$. What is the general solution of $d^2f/dx^2 = 0$? Solve the DEs $d^2f/dx^2 - 4f = 0$ and $d^2f/dx^2 + 2f = 0$, in each case subject to initial conditions $f = 2$, $df/dx = 0$ at $x = 0$.

3. Functions f_1, f_2 are the solutions of the DE $d^2f/dx^2 - xf = 0$, subject to initial conditions $f_1 = 1$, $df_1/dx = 0$ at $x = 0$, and $f_2 = 0$, $df_2/dx = 1$ at $x = 0$.

    If f is any solution of the DE, use the equations $d^2f/dx^2 = xf$, $d^2f_1/dx^2 = xf_1$, $d^2f_2/dx^2 = xf_2$, and the initial conditions for f_1 and f_2, to verify that

    (a) $\frac{d}{dx}\{f_1\frac{df_2}{dx} - f_2\frac{df_1}{dx}\} = 0$, hence $f_1\frac{df_2}{dx} - f_2\frac{df_1}{dx} = 1$;

    (b) the functions $f\frac{df_2}{dx} - f_2\frac{df}{dx}$, $-f\frac{df_1}{dx} + f_1\frac{df}{dx}$ both have zero derivative, hence $f\frac{df_2}{dx} - f_2\frac{df}{dx} = A$, $-f\frac{df_1}{dx} + f_1\frac{df}{dx} = B$, where A and B are constants;

    (c) $f(x) = Af_1(x) + Bf_2(x)$.

    Obtain the coefficients a_k in the Taylor series expansion $f_1(x) = \sum_0^\infty a_k x^k$ of the function f_1 about $x = 0$.

11.3 Constant coefficient differential equations

A constant coefficients DE of second order is a DE of the form

$$c_2 \frac{d^2f}{dx^2} + c_1 \frac{df}{dx} + c_0 f = p(x)$$

where the coefficients c_0, c_1, c_2 are all constants, with $c_2 \neq 0$. A constant coefficient DE of order n looks like

$$c_n \frac{d^n f}{dx^n} + c_{n-1} \frac{d^{n-1}f}{dx^{n-1}} + \dots + c_1 \frac{df}{dx} + c_0 f = p(x)$$

where all coefficients $c_0, c_1, \dots, c_{n-1}, c_n$ are constants, and $c_n > 0$. Most of the ideas underlying the solution of constant coefficients DEs are well illustrated by the case $n = 2$, so we shall look at this case first of all.

Consider first the homogeneous constant coefficients DE of second order, which takes the form

$$c_2 \frac{d^2f}{dx^2} + c_1 \frac{df}{dx} + c_0 f = 0$$

We have to find two independent solutions for f. Motivated by the example $d^2f/dx^2 - k^2 f = 0$ (see problem 2 of Exercises on 11.2), for which there are solutions $e^{\pm kx}$, it seems appropriate to look for exponential solutions.

Let us therefore try a solution $f(x) = e^{kx}$, where the constant k is to be determined. With $f = e^{kx}$, we have $df/dx = ke^{kx}$, $d^2f/dx^2 = k^2 e^{kx}$, and substituting into the homogeneous DE gives $(c_2 k^2 + c_1 k + c_0)e^{kx} = 0$. Certainly $e^{kx} \neq 0$, and the only possibility is $c_2 k^2 + c_1 k + c_0 = 0$. Hence we are led to a quadratic equation for the constant k. There are three possibilities to consider:

Possibility 1

Quadratic $c_2 k^2 + c_1 k + c_0 = 0$ has two real roots. This will happen provided $c_1^2 > 4c_0 c_2$. We then have two possible values of k, say $k = k_1$ and $k = k_2$, obtained by solving the quadratic equation. Hence $f_1(x) = e^{k_1 x}$ and $f_2(x) = e^{k_2 x}$ are two independent solutions, and the general solution of the homogeneous equation is $f(x) = A e^{k_1 x} + B e^{k_2 x}$.

Possibility 2

Quadratic $c_2 k^2 + c_1 k + c_0 = 0$ has equal roots. This will happen whenever $c_1^2 = 4c_0 c_2$.

Let the solution of the quadratic be $k = k_1$. We only have one value of k, as the root $k = k_1$ is repeated. The quadratic in this case factorizes as $c_2(k - k_1)^2$. Certainly $f(x) = e^{k_1 x}$ is a solution of the homogeneous DE, but we need a second solution to complete the determination of the general solution. Where is this second solution to come from? In fact, a little trial and error with exponential-type solutions in this case may eventually lead you to try the simplest combination of $e^{k_1 x}$ with a power of x, namely $f(x) = x e^{k_1 x}$.

If $f = xe^{k_1x}$, then $df/dx = (1 + k_1x)e^{k_1x}$ and $d^2f/dx^2 = (2k_1 + k_1^2x)e^{k_1x}$. Substituting into the homogeneous DE, we then have, on grouping together similar terms,

$$c_2\frac{d^2f}{dx^2} + c_1\frac{df}{dx} + c_0f = (c_2k_1^2 + c_1k_1 + c_0)xe^{k_1x} + (2c_2k_1 + c_1)e^{k_1x}$$

We need to show that the right-hand side is zero. Certainly the first term, involving xe^{k_1x}, is zero. The coefficient $(c_2k_1^2 + c_1k_1 + c_0)$ is zero simply because $k = k_1$ is our solution of the quadratic equation. What about the second term? Here we must show that $2c_2k_1 + c_1 = 0$. But from the solution of the quadratic with $c_1^2 = 4c_0c_2$ we have $k_1 = (-c_1 \pm \sqrt{0})/2c_2 = -c_1/2c_2$, implying that $2c_2k_1 + c_1 = 0$, so $f = xe^{k_1x}$ really is a solution of the homogeneous DE and can count as our second solution. The two solutions in this case are $f_1 = e^{k_1x}$, $f_2 = xe^{k_1x}$, and the general solution is the combination $f(x) = (A + Bx)e^{k_1x}$.

Possibility 3

Quadratic $c_2k^2 + c_1k + c_0 = 0$ has complex roots. This will happen whenever $c_1^2 < 4c_0c_2$.

From the formula for solution of the quadratic equation, or from a little algebra of complex numbers, you can verify that the two roots in this case occur as a pair, $k = \alpha \pm i\beta$, where α and β are real numbers. To determine the values of α and β, you will have to solve the quadratic equation (in fact, $\alpha = -c_1/2c_2$ and $\beta = \sqrt{4c_0c_2 - c_1^2}/2c_2$).

Once again, we do have two solutions in this case, namely $e^{(\alpha+i\beta)x}$ and $e^{(\alpha-i\beta)x}$. The problem is that these are **complex** functions of x. Nevertheless, do not throw them away — they are going to be very useful to us.

The point is, that if you have a complex solution, say $f(x) = u(x) + iv(x)$ where u and v are real, of a real DE such as $d^2f/dx^2 + df/dx + f = 0$, then both u and v must separately satisfy the DE. You can see this, because on substituting $f = u + iv$ into the DE and collecting together those terms with i and those terms without i, you will find

$$\left(\frac{d^2u}{dx^2} + \frac{du}{dx} + u\right) + i\left(\frac{d^2v}{dx^2} + \frac{dv}{dx} + v\right) = 0$$

Such an expression can only be zero if both bracketed functions are separately zero, and this means both u and v satisfying the DE. The same argument, which by the way once more invokes linearity of differentiation, applies to any (real) homogeneous DE, whether constant coefficients or not.

In the present context, this implies that we can obtain **real** solutions by splitting $e^{(\alpha+i\beta)x}$ or $e^{(\alpha-i\beta)x}$ into a real part and a complex part. To do this, we must refer back to Section 7.4, where I dealt briefly with the algebra of complex numbers and exponentials. Writing $e^{(\alpha+i\beta)x}$ as a product $e^{\alpha x}.e^{i\beta x}$ and using De Moivre's Theorem (equation (7.11)), we have

$$e^{(\alpha+i\beta)x} = e^{\alpha x}(\cos \beta x + i \sin \beta x)$$

Hence both the real part, $e^{\alpha x} \cos \beta x$, and the complex part, $e^{\alpha x} \sin \beta x$ must separately satisfy the homogeneous DE. We can use the same argument with

$e^{(\alpha - i\beta)x}$, but the two solutions in that case turn out to be $e^{\alpha x} \cos \beta x$ and $-e^{\alpha x} \sin \beta x$ and are not essentially different. It follows, in the case of the quadratic having complex roots $\alpha \pm i\beta$, that once again two (real) solutions are obtained. They are $f_1(x) = e^{\alpha x} \cos \beta x$ and $f_2(x) = e^{\alpha x} \sin \beta x$, and the general solution in this case is $f(x) = e^{\alpha x}(A \cos \beta x + B \sin \beta x)$.

The situation for homogeneous constant coefficients DEs of order n is very similar. In the case of these higher order DEs, one has a polynomial equation of degree n to solve for the index k in the exponential. This equation will always have n (real or complex) roots, but they need not all be distinct. Any complex roots will occur in pairs $k = \alpha \pm i\beta$, and the same analysis as in the second order case will produce two real solutions for each complex pair. Repeated roots for k may have multiplicity greater than 2, but the same principle applies as before. For example if a root $k = k_1$ occurs four times, i.e. a root of multiplicity 4, then further solutions can be obtained on multiplying by powers of x, giving four solutions $f_1 = e^{k_1 x}$, $f_2 = xe^{k_1 x}$, $f_3 = x^2 e^{k_1 x}$ and $f_4 = x^3 e^{k_1 x}$. All this can be summed up in the following strategy for solution of such equations.

SOLUTION
Such a DE looks like

$$c_n \frac{d^n f}{dx^n} + c_{n-1} \frac{d^{n-1} f}{dx^{n-1}} + \dots + c_1 \frac{df}{dx} + c_0 f = 0$$

Try exponentials of the form $f = e^{kx}$, leading to the polynomial equation for k:

$$c_n k^n + c_{n-1} k^{n-1} + \dots + c_1 k + c_0 = 0$$

Each real root for k leads to a corresponding solution $f(x) = e^{kx}$. A repeated root of multiplicity N leads to the N solutions $f(x) = e^{kx}, xe^{kx}, x^2 e^{kx}, \dots, x^{N-1} e^{kx}$.

Complex roots occur in pairs, and each complex pair $k = \alpha \pm i\beta$ leads to a corresponding pair of solutions

$$f(x) = e^{\alpha x} \cos \beta x, \quad e^{\alpha x} \sin \beta x$$

Hence n (real) solutions for $f(x)$ are obtained, corresponding to the order n of the homogeneous DE. If these solutions are labelled $f_1, f_2, \dots, f_n$, the general solution of the homogeneous DE is $f(x) = A_1 f_1(x) + A_2 f_2(x) + \dots + A_n f_n(x)$.

This strategy enables us to solve any homogeneous constant coefficients DE, provided the roots of the polynomial equation can be determined. (Even if they cannot, we at least know the **form** that the general solution will take, and the polynomial roots can be approximated numerically.) The following examples illustrate the procedure.

Example 7

I have grouped together several examples of this genre under this heading.

(1) $2\frac{d^2 f}{dx^2} + \frac{df}{dx} - f = 0$: $f = e^{kx}$ leads to $2k^2 + k - 1 = 0$. Hence $k = \frac{1}{2}$ or $k = -1$. General solution: $f = Ae^{\frac{1}{2}x} + Be^{-x}$.

(2) $\frac{d^2 f}{dx^2} - 4\frac{df}{dx} + 4f = 0$: here $k^2 - 4k + 4 = 0$, or $(k-2)^2 = 0$, so $k = 2$ is a double root. We have two solutions, e^{2x}, xe^{2x}. General solution: $(A + Bx)e^{2x}$.

(3) $\frac{d^2y}{dt^2} + \frac{dy}{dt} + y = 0$: with $y(t) = e^{kt}$ we have $k^2 + k + 1 = 0$. Complex roots $k = (-1 \pm \sqrt{-3})/2 = -\frac{1}{2} \pm (\sqrt{3}/2)i$. Complex solutions are $y(t) = e^{(-\frac{1}{2} \pm \frac{i\sqrt{3}}{2})t}$ and the real solutions are $e^{-\frac{1}{2}t} \cos(\sqrt{3}/2)t$, $e^{-\frac{1}{2}t} \sin(\sqrt{3}/2)t$. General solution:

$$y(t) = e^{-\frac{1}{2}t}\left(A \cos \frac{\sqrt{3}}{2}t + B \sin \frac{\sqrt{3}}{2}t\right)$$

(4) $\frac{d^4u}{d\theta^4} - u = 0$: here $u(\theta) = e^{k\theta}$ gives $k^4 - 1 = 0$. Hence $(k^2 - 1)(k^2 + 1) = 0$, with the four roots $k = \pm 1$, $\pm i$. The two real solutions are e^{θ}, $e^{-\theta}$ and the two complex solutions are $e^{\pm i\theta}$. Since $e^{i\theta} = \cos\theta + i\sin\theta$, a further two solutions are $\cos\theta$ and $\sin\theta$. General solution: $u(\theta) = Ae^{\theta} + Be^{-\theta} + C\cos\theta + D\sin\theta$.

(5) $\frac{d^3y}{dx^3} + 3\frac{d^2y}{dx^2} + 3\frac{dy}{dx} + y = 0$: $y = e^{kx}$ leads to $(k+1)^3 = 0$. The triple root $k = -1$ gives rise to three solutions, e^{-x}, xe^{-x}, x^2e^{-x}. General solution: $y = e^{-x}(A + Bx + Cx^2)$.

To write down the general solution of the **inhomogeneous** constant coefficients DE, we must find just one particular solution to add to the general solution of the homogeneous DE. Although there is no universal, infallible general method by which such solutions can be found, a particular solution can **always** be determined if its general form is known. For example, if you know there is a solution which is a quadratic polynomial, it is only necessary to substitute $f = ax^2 + bx + c$ into the DE and use a little algebra to evaluate the coefficients. In practice, one looks for a solution of a particular form, dependent on the nature of the right-hand side function $p(x)$. A good rule of thumb is to try solutions which are based on some combination of the function $p(x)$ itself and functions obtained by substituting $p(x)$ into the left-hand side of the DE. Here are three typical examples to illustrate the method.

Example 8

We consider the DE $d^2f/dx^2 + f = 1 + x^2$. In view of the right-hand side function, an obvious choice for a particular solution would be a second degree polynomial.

With $f = ax^2 + bx + c$, the left-hand side becomes $ax^2 + bx + (c + 2a)$. Thinking of the right-hand side as $x^2 + 0.x + 1$, we need $a = 1$, $b = 0$ and $c + 2a = 1$. That is, $a = 1$, $b = 0$ and $c = -1$. The particular solution is $x^2 - 1$. Since the homogeneous equation has solution $A\cos x + B\sin x$, the general solution of the DE is $f(x) = x^2 - 1 + A\cos x + B\sin x$.

Example 9

Let the function f satisfy $d^2f/dx^2 - f = e^{-2x}$. An obvious trial function for f is a multiple of e^{-2x}. Substituting $f = ae^{-2x}$, we have $3ae^{-2x} = e^{-2x}$, so that $a = \frac{1}{3}$.

The particular solution is $\frac{1}{3}e^{-2x}$, and the general solution is

$$f(x) = \frac{1}{3}e^{-2x} + Ae^x + Be^{-x}$$

❋ *Example 10*

We consider the DE $d^2y/dt^2 + dy/dt + y = \cos t$, with $y(0) = 0$, $y'(0) = 0$. We might try a multiple of $\cos t$ for a particular solution. However, the substitution of $\cos t$ into the left-hand side of the DE generates a sine term as well, so why not try a combination of a cosine and a sine, say $y(t) = a\cos t + b\sin t$. Then $d^2y/dt^2 + dy/dt + y = -a\sin t + b\cos t$, where we need $a = 0$ and $b = 1$, so the solution in the end does not include a cosine at all! It is $y_0(t) = \sin t$.

For the homogeneous equation, already treated in Example 7, we have the solution $e^{-\frac{1}{2}t}(A\cos(\sqrt{3}/2)t + B\sin(\sqrt{3}/2)t)$. Hence the general solution is $y(t) = \sin t + e^{-\frac{1}{2}t}(A\cos(\sqrt{3}/2)t + B\sin(\sqrt{3}/2)t)$.

Applying the initial conditions, $y(0) = 0$ gives $A = 0$, and $y'(0) = 0$ gives $1 - \frac{1}{2}A + (\sqrt{3}/2)B = 0$. Hence $A = 0$, $B = -2/\sqrt{3}$, and we are left with the solution $y(t) = \sin t - (2/\sqrt{3})e^{-\frac{1}{2}t}\sin(\sqrt{3}/2)t$.

EXERCISES ON 11.3

1. With the notation D for d/dx, the differential equation $2\,d^2f/dx^2 + df/dx - f = 0$ may be written as $(2D - 1)(D + 1)f = 0$. Show that a function g can be defined such that $df/dx + f = g$, $2\,dg/dx - g = 0$. Write down the general solution for g, and hence solve the first order linear DE for f to show that $f(x) = Ae^{\frac{1}{2}x} + Be^{-x}$, where A and B are constants.
2. Write the DE $d^2f/dx^2 - 4\,df/dx + 4f = 0$ in the form $(D - 2)^2f = 0$, and use the method of problem 1 to show that the general solution is $f(x) = (A + Bx)e^{2x}$.
3. The function $f(x)$ satisfies the DE $4x^2\,d^2f/dx^2 + f = 0$ ($x > 0$). A function $g(t)$ is defined by $g(t) = f(e^t)$. Use the chain rule to show that $dg/dt = e^tf'(e^t)$, $d^2g/dt^2 = e^{2t}f''(e^t) + dg/dt$, and hence show that $g(t)$ satisfies the DE $4\,d^2g/dt^2 - 4\,dg/dt + g = 0$. Find the general solution for $g(t)$, and hence solve the DE for f.
4. Find the general solution of each of the DEs:

$$\frac{d^2f}{dx^2} + f = x^4, \qquad \frac{d^2f}{dx^2} - \frac{df}{dx} = \sin x$$

Summary

There is no general method which will provide us with the solution (or even **a** solution) in closed form of every linear DE that we are ever likely to meet. In the case of linear first order DEs, there **is** a general method of solution, which makes use of integrating factors. An integrating factor for the first order DE $df/dx + p_0(x)f = p(x)$ is $P(x) = \exp \int p_0(x)\,dx$. On multiplying through by this integrating factor, the DE becomes

$$\frac{d}{dx}(P(x)f) = p(x)P(x)$$

and may be solved by integration.

For a linear DE of arbitrary order n, the general solution is always of the form $f(x) = f_0(x) + \sum_1^n A_k f_k(x)$, where f_0 is a particular solution of the (inhomogeneous) DE, and the f_k are n independent solutions of the homogeneous DE.

Table 11.1 Trial functions for solution of constant coefficient DEs.

Right-hand side function $p(x)$ equals	Trial function $f_0(x)$ equals
e^{cx}	ae^{cx} [or $ax^r e^{cx}$, if c is a root of multiplicity r of the polynomial equation for k]
$\cos cx$ or $\sin cx$	$a \cos cx + b \sin cx$
Polynomial of degree N	Polynomial of degree N
$e^{mx} \cos cx$ or $e^{mx} \sin cx$	$e^{mx}(a \cos cx + b \sin cx)$

In the case of a constant coefficients linear DE of order n, the homogeneous equation has solutions of the form e^{kx}, where the possible values of k are solutions of a polynomial equation of degree n, obtained by substituting $f = e^{kx}$ into the homogeneous DE. The required n solutions of the homogeneous DE can be constructed in this way, provided appropriate modifications are made to take into account possible complex roots or repeated roots of the polynomial equation.

A particular solution of the inhomogeneous DE for a constant coefficient equation can often be found by substituting a suitable trial function into the DE. The choice of trial function will depend on the form of the right-hand side function $p(x)$ of the DE. Some suggestions for trial functions are given in Table 11.1.

FURTHER EXERCISES

1. Solve, by use of an integrating factor, the DE $x\,df/dx + 3f = 5x^2$ $(x > 0)$, subject to the condition $f(1) = 2$.
2. Find the general solution, for $x > 0$, of the DE $x\,df/dx + (2x - 1)f = x^2$.
3. Verify by differentiation that the function

$$y(x) = x\left(c + \int_1^x \frac{h(t)}{t^2}\,dt\right)$$

satisfies the DE $x\,dy/dx - y = h(x)$, subject to the condition $y(1) = c$, where h is a given function. Hence solve the DE $x\,dy/dx - y = x$, subject to the condition $y(1) = 1$.
4. Solve the DE $d^2f/dx^2 - 4\,df/dx - 12f = e^{-3x}$, subject to initial conditions $f = 0$ and $df/dx = -1$ at $x = 0$, and find the general solution of the DE

$$\frac{d^2f}{dx^2} - 4\frac{df}{dx} - 12f = 1 - e^{-3x}$$

5. Solve the DE $d^2f/dx^2 + 2\,df/dx + 2f = 0$, subject to the conditions $f(0) = 0$, $f'(0) = 1$; sketch the graph of this solution, for positive values of x.
6. A function f is defined in terms of a given function $h(x)$ by $f(x) = \frac{1}{2}e^x \int_0^x e^{-t}h(t)\,dt - \frac{1}{2}e^{-x} \int_0^x e^t h(t)\,dt$. Verify that

$$\frac{df(x)}{dx} = \frac{1}{2}e^x \int_0^x e^{-t}h(t)\,dt + \frac{1}{2}e^{-x} \int_0^x e^t h(t)\,dt$$

Write down an expression for $d^2f(x)/dx^2$, and verify that the function f satisfies the DE $d^2f/dx^2 - f = h$, subject to the conditions that $f = 0$ and $df/dx = 0$ at $x = 0$. Use this result to solve the DEs $d^2f/dx^2 - f = e^x$ and $d^2f/dx^2 - f = xe^x$, in each case subject to the conditions $f = 0$, $df/dx = 0$ at $x = 0$.

7. Verify that the function y, given by

$$y(x) = \sin x \int_0^x h(t) \cos t \, dt - \cos x \int_0^x h(t) \sin t \, dt$$

satisfies the DE $d^2y/dx^2 + y = h(x)$, with initial conditions $y = 0$, $dy/dx = 0$ at $x = 0$. Hence or otherwise obtain the general solution of $d^2y/dx^2 + y = \cos x$.

8. Find two values of the constant c such that x^c is a solution of the DE

$$x^2 \frac{d^2f}{dx^2} + x \frac{df}{dx} - 4f = 0, \quad \text{for } x > 0$$

Hence solve the DE, subject to the conditions $f = 0$, $df/dx = 1$ at $x = 1$.

9. The function $f(t)$ satisfies the DE

$$t \frac{d^2f}{dt^2} + (t + 1) \frac{df}{dt} + f = 4t \quad (t > 0)$$

with initial conditions $f = 1$, $df/dt = 1$ at $t = 1$. Verify by successive differentiation that $f(t)$ satisfies

$$t \frac{d^3f}{dt^3} + (t + 2) \frac{d^2f}{dt^2} + 2 \frac{df}{dt} = 4 \quad \text{and} \quad t \frac{d^4f}{dt^4} + (t + 3) \frac{d^3f}{dt^3} + 3 \frac{d^2f}{dt^2} = 0$$

How does this sequence of equations continue? Verify that $d^n f/dt^n = (-1)^n$ at $t = 1$, for all $n \geq 2$, and hence write down a Taylor series for f in powers of $(t - 1)$.

10. Verify that the second order DE in exercise 9 can be written as

$$\frac{d}{dt} \left\{ t \frac{df}{dt} + tf \right\} = 4t$$

Hence obtain a first order DE satisfied by f, of the form $df/dt + f = p(t)$, where the function p is to be determined. Solve this first order DE for f, and show that your solution agrees with the power series solution of Exercise 9.

$12 \bullet$ Looking Back and Looking Forward

In this final chapter, I would like briefly to review some of the ideas and methods which we have met in previous chapters, and then look ahead to some of the ways in which you may wish to follow up some of these ideas and methods by pursuing the subject further.

This book is an introductory test of calculus at university level. While the level is appropriate to a first year module, much of the material has been presented in a way which will help to provide a springboard for courses at a more advanced level which can be followed later.

12.1 Looking back

In taking a backward glance at some of the material we have covered, I will do no more than indicate, from each chapter, one or two areas where I have tried to explain an important idea or describe a new method. I hope that the following compilation will remind you of some of the main points, and will help guide your revision. We begin with Chapter 2, which followed the Introduction.

In Chapter 2, I identified the idea of function as being central to the subject. A function is a rule for sending numbers to numbers, and to define a function completely you need to specify not only this rule (which must send a given real number to just one real number) but also the **domain** and **range** of the function. Though this is not a book on analysis, we looked briefly at the idea of a **continuous function**, and of a **limit**.

In Chapter 3 ways of combining functions were looked at, using the usual algebraic operations of addition, multiplication, and so on, but also defining the **composition** of two functions and the notion of an **inverse**. Not every function has an inverse, but an **injective** function will always have an inverse, and if a function is not injective you can restrict its domain so that it becomes so.

In Chapter 4 the idea of the derivative or rate of change of a function was considered. The derivative may be interpreted geometrically as the gradient of the graph of a function. We saw how to write down a formal expression for a derivative as a limit, and how to evaluate this limit 'from first principles'. This led us to the equation of the tangent to the graph at a point; the tangent may be regarded as locally the best fit of the graph by a straight line. We discussed the interpretation of the derivative in kinematics in terms of speed and velocity.

Chapter 5 was mainly about the mechanics of differentiation, or how to **evaluate** the derivative. We saw how to differentiate some of the standard functions, and found rules for differentiating products, quotients, powers and inverse functions. You should also have become familiar with the use of the chain rule. Some of the standard functions are best defined by power series. Provided we are careful to remain within the radius of convergence, the differentiation of power series can be carried out without difficulty. We saw that differentiation is a linear operation.

In Chapter 6, I introduced the notation for higher order derivatives, and showed how to write down the Taylor series for a function about a point. The first derivative provides information on the existence of critical points, and of points of increase or decrease, which can be used to determine the position and nature of any maxima/minima. These ideas were partly directed to the developing of a general strategy for analysing functions and sketching their graphs.

Chapter 7 began with the exponential function, which was defined by means of its Taylor series. Most of the special functions of calculus, including the logarithm, powers, and hyperbolic function, may be expressed in terms of the exponential. We explored their main properties, and found that by the use of complex numbers even the sine and cosine function, which appear to be quite different, are close cousins of the exponential.

If the derivative of $F(x)$ is $f(x)$, say for x in some interval, we say that $F(x)$ is an **antiderivative** of $f(x)$. Any other antiderivative of $f(x)$ (on the same interval) will differ from $F(x)$ by a constant. Antidifferentiation is the inverse operation to differentiation, and was the subject of Chapter 8. The general antiderivative of a function $f(x)$ is written as $\int f(x)\,dx + c$, and is called the **indefinite integral** of $f(x)$. Antidifferentiation is fundamental to many areas of mathematics, including the theory of DEs, and has wide applications to dynamics. Among the methods of integration described in this chapter were integration by parts and integration by change of variable, as well as the use of partial fractions.

In Chapter 9, we went on to consider the definite integral. The indefinite integral is an antiderivative, but the definite integral is an area. There are many other interpretations of the definite integral, and we looked at some of them, as well as some applications. The definite and indefinite integral are linked by the identity

$$\frac{d}{dx} \int_a^x f(t)\,dt = f(x)$$

which holds for any function continuous on an interval, and asserts that $\int_a^x f(t)\,dt$ is an antiderivative of $f(x)$, for any constant a. From this chapter, you will have a good idea of how to evaluate many of the definite integrals which you will come across, including examples of improper integrals, for which the interval of integration may be infinite.

The last two chapters of the book have as their subject one of the main areas of application of the ideas and methods of calculus, namely the theory and application of DEs. Chapter 10 dealt principally with first order equations, that is equations of the form $df/dx = Q(x, f)$, where Q is a given function. Where the DE can be solved 'exactly' (as for example in the case of separable DEs), I have explained how this can be done; but it is also important to realize that in very many cases of importance in applications no precise analytic solution can be found, or it can only be found with difficulty. In that case, numerical and other methods can be used, and I described a number of techniques including graphical methods, series solutions and iterations. These methods are underpinned by a theoretical framework in which one can show quite generally that there is one and only one solution subject to a given initial condition $f(x_0) = y_0$.

In the more general situation of a DE of order n, n initial conditions are needed to determine a solution. In dynamics, DEs of order 2 lead to a unique solution of the equations of motion, for any given initial state of the system.

Chapter 11 exploits the linearity of the operation of differentiation in treating the important category of **linear** DEs. Any linear DE of order n can be solved completely provided a single particular solution of the DE can be found, together with n independent solutions of the corresponding homogeneous DE. In the case of a first order equation, the exact solution can be obtained. In the more general situation of order n, a number of special cases can be handled, including constant coefficients equations which are of importance in many applications.

12.2 Signposts

I very much hope that what you have found in this book will have stimulated your interest sufficiently for you to wish to follow up its contents by further study. At Hull, we offer a number of modules, both for mathematicians and for the principal users of mathematics, which develop and extend much of the work which we have begun here. It is my aim to have provided a firm foundation on which to build, wherever and however you may choose to continue.

Here are some signposts, which may help to guide you on your way. Each signpost points from a chapter of this book.

2:$\Rightarrow$:For a more formal approach to limits and continuity, see any text on analysis.

3:$\Rightarrow$:A good way to understand how a function is put together by a sequence of elementary operations is through use of a flow diagram. These flow diagrams can also be used as the basis of computer programs to evaluate functions.

4:$\Rightarrow$:The formal definition of derivative as a limit may be extended to apply to functions of a complex variable z. Thus $f'(z) = \lim_{h \to 0}[f(z+h) - f(z)]/h$, where now the function f, as well as z and h, are all complex. A function $f(z)$ such that $f'(z)$ exists for all z in some open domain of the complex plane is said to be **analytic** in that domain. The theory of analytic functions is one of the most elegant and widely applicable areas of modern analysis.

5:$\Rightarrow$:The theory of differentiation may be extended to functions of more than one variable. Applied to functions $f(x, y)$ of two variables, for example, one can define a kind of vector differential operator ∇, such that the two components $\nabla = (\partial/\partial x, \partial/\partial y)$ of ∇ correspond to differentiation with respect to x and y, respectively.

6:$\Rightarrow$:The search for maxima/minima of functions may also be extended to functions of several variables, and leads to the theory of optimization.

7:$\Rightarrow$:Most of the special functions of calculus can be defined also as functions of a complex variable z. They then become analytic in some suitable domain (see 4).

8:$\Rightarrow$:The antiderivative may be regarded as an inverse operation to differentiation. More generally, it is often possible to define an inverse of second order or higher order differential operators. This leads to the theory of integral operators, with close links to the subject of integral equations, and to the modern theory of functional analysis.

9:⇒:Many different kinds of integral may be defined for functions in two or more dimensions. For example, one may integrate over curves, or surfaces, or volumes. These integrals are important in the analysis of vector fields, with many applications to the physics and engineering of continuous media.

10:⇒:A set of coupled differential equations, called Lagrange's equations, can be used as the basis for an approach to dynamics. This leads to the theory of analytic dynamics, which can be applied to all conservative dynamical systems, and provides a systematic method of writing down the equations of motion.

11:⇒:Just as every first order linear DE may be solved by use of an integrating factor, so for example every **second** order DE, on multiplying by an integrating factor, may be expressed as the derivative of a first order DE. However, the integrating factor is itself determined as a solution of a second order DE, which may be as difficult to solve as the original equation. Nevertheless, the theory of integrating factors for higher order DEs is a useful and important extension of the theory covered in this book.

Solutions

There are exercises at the end of each section, and at the end of each chapter. Solutions to exercises at the end of each chapter are provided below.

Chapter 2

1. If $f(x) = \sqrt{9 - x^2}$, then the graph is described by the equation $y = \sqrt{9 - x^2}$. Hence $x^2 + y^2 = 9$. This is the equation of a circle, or rather a semicircle since $y \geq 0$.

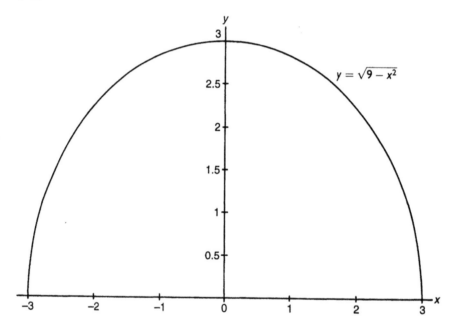

$f(-2) = \sqrt{5}, f(0) = 3, f(1) = \sqrt{8} = 2\sqrt{2}$.
Domain is the interval $[-3, 3]$.
Range is the interval $[0, 3]$.

2. See opposite.

3. For N even, $(-1)^N = 1$, and $y(t) = t^N$ satisfies $y(-t) = (-t)^N = (-1)^N t^N = t^N$. So $y(-t) = y(t)$, and y is an even function.
Similarly, if N is odd, then $(-1)^N = -1$, and $y(-t) = (-1)^N y(t) = -y(t)$, so that y is an odd function.
The functions defined by $y(t) = t^2 + t^3$, and $y(t) = |t| + t$, respectively, are neither even nor odd, since in the first case $y(-t) = t^2 - t^3$ and in the second $y(-t) = |t| - t$; in neither example do we have $y(-t) = y(t)$ or $y(-t) = -y(t)$.

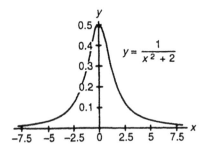

$$y = \frac{1}{x^2 + 2}$$

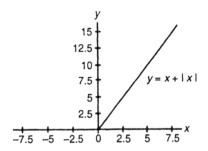

$y = x + |x|$

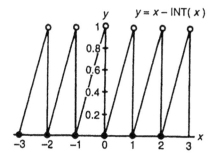

$y = x - \text{INT}(x)$

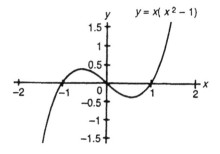

$y = x(x^2 - 1)$

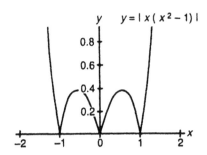

$y = |x(x^2 - 1)|$

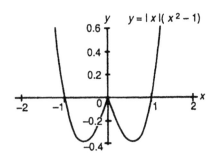

$y = |x|(x^2 - 1)$

If $y(t)$ is written as the sum of $y_1(t)$ and $y_2(t)$, where $y_1(t) = \frac{y(t)+y(-t)}{2}$ and $y_2(t) = \frac{y(t)-y(-t)}{2}$, then $y_1(-t) = \frac{y(-t)+y(+t)}{2} = y_1(t)$, and $y_2(-t) = \frac{y(-t)-y(+t)}{2} = -y_2(t)$.

Hence y_1 is an even function, and y_2 is odd.

Chapter 3

1. $\dfrac{1+x}{1+2x} \geq 0$ if **either** $1+x \geq 0$ and $1+2x > 0$
   or $1+x \leq 0$ and $1+2x < 0$.

The first case will occur only for $x > -\frac{1}{2}$.
The second case will occur only for $x \leq -1$.
Hence $(1+x)/(1+2x) \geq 0$ for $x \leq -1$ and for $x > -\frac{1}{2}$; that is, in interval notation, for $x \in (-\infty, -1] \cup (-\frac{1}{2}, \infty)$.

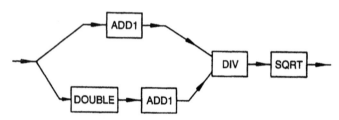

Domain is $(-\infty, -1] \cup (-\frac{1}{2}, \infty)$.
Range is $[0, \frac{1}{\sqrt{2}}) \cup (\frac{1}{\sqrt{2}}, \infty)$.

2. $x^2 + 2x + 4 = (x+1)^2 + 1 + 1 + 1$ and is the result of the successive operations ADD1, SQUARE, ADD1, ADD1, ADD1.
   So this function is $g \circ g \circ g \circ f \circ g$ with $f =$ 'square' and $g =$ 'add1'.

3. $f_1 \circ f_2$ is the function f given by $f(x) = \sqrt{x^2}$. Note $\sqrt{\bullet}$ denotes positive square root; so $f(x) = |x|$ is the absolute value or modulus function.

4. $f(x) = 1/x$; $f^{-1}(x) = 1/x$; domain and range of f are $\mathbb{R}\setminus\{0\}$.
   $f(x) = (x^3+1)/(x^3-1)$; $f^{-1}(x) = \sqrt[3]{(x+1)/(x-1)}$; domain and range of f are $\mathbb{R}\setminus\{1\}$; $f(x) = 2\sqrt{x} + 1$; $f^{-1}(x) = (x-1)^2/4$ for $x \geq 1$; domain of f is $[0, \infty)$, range of f is $[1, \infty)$.
   In each case, the domain of f^{-1} is the same as the range of f, and the range of f^{-1} is the same as the domain of f.

5. The identity holds for all $x \in [-1, 1]$.
   Here $y_1 = \sin^{-1} x$ is defined by $\sin y_1 = x$ and $-\pi/2 \leq y_1 \leq \pi/2$. Hence $y_2 = (\pi/2) - y_1$ satisfies $\cos y_2 (= \sin y_1) = x$, and $0 \leq y_2 \leq \pi$. It follows that $y_2 = \cos^{-1} x$, and we have $y_1 + y_2 = \sin^{-1} x + \cos^{-1} x = \pi/2$. For $x \in \mathbb{R}\setminus(-1, 1)$, define $y = \sec^{-1} x$ by $x = \sec y$ and $0 \leq y \leq \pi$. Then $1/x = \cos y$ with $1/x \in [-1, 1]$, hence $y = \cos^{-1}(1/x)$. So $\sec^{-1} x = \cos^{-1}(1/x)$.

6. (a) $f(t) = \frac{1+t}{t}$, $(f \circ f)(t) = \frac{1+2t}{1+t}$, $(f \circ f \circ f)(t) = \frac{2+3t}{1+2t}$, $(f \circ f \circ f \circ f)(t) = \frac{3+5t}{2+3t}$, ...

   Each function on the right-hand side is of the form $(a + bt)/(c + dt)$. Successive values of a, b, c and d are related to the so-called Fibonacci sequence, for which each term is the sum of the two previous terms. For example, successive values of a are $1, 1, 2, 3, 5, 8, 13, \ldots$ which form the Fibonacci sequence itself. If a_n denotes the nth number from the Fibonacci sequence, then the nth function will be $(a_n + a_{n+1}t)/(a_{n-1} + a_n t)$.

(b) $x_1 = 1$, $x_2 = 2$, $x_3 = 3/2$, $x_4 = 5/3, \ldots$.
If $x_{n+1} = 1 + (1/x_n)$, and $x_n \to x$, then $x = 1 + (1/x)$, or $x^2 - x - 1 = 0$. Since clearly $x > 0$, we must have $x = (1 + \sqrt{5})/2$. (This is the so-called 'Golden ratio', and has the value $x = 1.618\ldots$)
The relation between (a) and (b) comes from the fact that in both instances we are iterating the function $1 + (1/t)$. The connection between the sequence $\{x_n\}$ and the Fibonacci sequence is that $x_n = a_{n+1}/a_n$.

Chapter 4

1. Differentiability at $x = x_0$ implies that $[f(x_0 + h) - f(x_0)]/h \to f'(x_0)$ as $h \to 0$. Hence, as $h \to 0$, $f(x_0 + h) - f(x_0) = h\{[f(x_0 + h) - f(x_0)]/h\}$ approaches zero. This is the same as the continuity of f at $x = x_0$. $f(x) = |x|$ is continuous but not differentiable at $x = 0$.

2. $\frac{\cos x_2 - \cos x_1}{x_2 - x_1} = 2 \sin\left(\frac{x_1 - x_2}{2}\right) \sin\left(\frac{x_1 + x_2}{2}\right)/(x_2 - x_1) = -\{\sin\left(\frac{x_1 - x_2}{2}\right)/\left(\frac{x_1 - x_2}{2}\right)\} \sin\left(\frac{x_1 + x_2}{2}\right)$.
Now let $\theta = (x_1 - x_2)/2$ and let $x_1, x_2 \to x$. In the limit, $\sin \theta/\theta \to 1$ and $\sin((x_1 + x_2)/2) \to \sin x$. Hence $\frac{\cos x_2 - \cos x_1}{x_2 - x_1} \to -\sin x$ which is the derivative of $\cos x$.

3. $\frac{g(x+h) - g(x)}{h} = \frac{(x+h)f(x+h) - xf(x)}{h} = \frac{x(f(x+h) - f(x)) + hf(x+h)}{h} = x\{\frac{f(x+h) - f(x)}{h}\} + f(x + h)$.
As $h \to 0$, $g'(x) = \lim_{h \to 0} \frac{g(x+h) - g(x)}{h} = \lim_{h \to 0} x\{\frac{f(x+h) - f(x)}{h}\} + \lim_{h \to 0} f(x + h) = xf'(x) + g(x)$.
The derivative of x is 1. Now apply the above result to $x.x$; the derivative of $x.x$ is $x.1 + x = 2x$. The derivative of $x.x^2 = x.2x + x^2 = 3x^2$. The derivative of $x.x^3$ is $x.3x^2 + x^3 = 4x^3$, and so on. So the derivative of x^n is nx^{n-1}.

4. If $u(t) = t^2 + 2t$ then $\frac{u(t+h) - u(t)}{h} = \frac{(t+h)^2 + 2(t+h) - t^2 - 2t}{h} = 2t + 2 + h \to 2t + 2$ as $h \to 0$; so $u'(t) = 2t + 2$. If $u(t) = 1/t^2$ then $\frac{u(t+h) - u(t)}{h} = (\frac{1}{(t+h)^2} - \frac{1}{t^2})/h = -\frac{(2t+h)}{(t+h)^2.t^2} \to -\frac{2}{t^3}$ as $h \to 0$; so $u'(t) = -2/t^3$.

5. If $y = x^2 + 2x$ then $y' = 2x + 2$. At $x = 1$, $y = 3$, we have $y' = 4$. Tangent $y = 4x + c$; $y(1) = 3 \Rightarrow c = -1$. So tangent is $y = 4x - 1$.
If $y = 1/x^2$, then $y' = -2/x^3$. At $x = 2$, $y = \frac{1}{4}$, we have $y' = -\frac{1}{4}$. Tangent $y = -\frac{1}{4}x + \frac{3}{4}$.

6. Since $|h| < x$, we have $h^2/x^2 < 1$ and $h/x > -1$; so $1 + (h/2x) - (h^2/2x^2) > 1 - \frac{1}{2} - \frac{1}{2} = 0$; hence $1 + (h/2x) - (h^2/2x^2) > 0$. Also $1 + (h/x) > 1 - 1 = 0$, and $1 + (h/2x) > 1 - \frac{1}{2} = \frac{1}{2} > 0$.
$(1 + (h/2x) - (h^2/2x^2))^2 < (1 + (h/x))$ is equivalent, on evaluating the square on the left-hand side and subtracting, to $\frac{3}{4} + (h/2x) - (h^2/4x^2) > 0$, where we have cancelled h^2/x^2. This inequality holds because $h/2x > -\frac{1}{2}$ and $h^2/4x^2 < \frac{1}{4}$,

so that $\frac{3}{4} + (h/2x) - (h^2/4x^2) > \frac{3}{4} - \frac{1}{2} - \frac{1}{4} = 0$; and $(1 + (h/x)) < (1 + (h/2x))^2$ is equivalent to $h^2/4x^2 > 0$. The inequalities $1 + (h/2x) - (h^2/2x^2) < \sqrt{1 + (h/x)} < 1 + (h/2x)$ follow from the two previous inequalities stated in the exercise, on taking square roots. If $f(x) = \sqrt{x}$, we have $\frac{f(x+h)-f(x)}{h} = \frac{\sqrt{x+h}-\sqrt{x}}{h} = \frac{\sqrt{x}}{h}\left\{\sqrt{1 + \frac{h}{x}} - 1\right\}$, which from the last inequalities lies between $(1/2\sqrt{x}) - (h/2x^{3/2})$ and $1/2\sqrt{x}$. Taking the limit $h \to 0$, we have $f'(x) = 1/2\sqrt{x}$.

Chapter 5

1. $\frac{d}{dx}(x^2 + x^4) = 2x + 4x^3$, $\quad \frac{d}{dx}(2x + \sqrt{x}) = 2 + \frac{1}{2\sqrt{x}}$, $\quad \frac{d}{dx}(x^2 \sin x) = x^2 \cos x + 2x \sin x$, $\quad \frac{d}{dx}\sin^2 x = 2 \sin x \cos x$, $\quad \frac{d}{dx}(\sqrt{x}\cos x) = -\sqrt{x}\sin x + \frac{1}{2\sqrt{x}}\cos x$, $\frac{d}{dx}\frac{x^2}{1+x^4} = \frac{2x(1-x^4)}{(1+x^4)^2}$, $\frac{d}{dx}\frac{1-x^2}{1+x+x^2} = -\frac{(1+4x+x^2)}{(1+x+x^2)^2}$, $\frac{d}{dx}\frac{\sin x}{1+\cos^2 x} = \frac{3\cos x - \cos^3 x}{(1+\cos^2 x)^2}$, $\frac{d}{dx}\sin(x^2 + 1) = 2x\cos(x^2 + 1)$, $\quad \frac{d}{dx}x\cos(x^2 + 1) = -2x^2 \sin(x^2 + 1) + \cos(x^2 + 1)$, $\frac{d}{dx}\frac{\sin 2x}{x} = \frac{2x\cos 2x - \sin 2x}{x^2}$, $\quad \frac{d}{dx}(1 + \cos 2x)^3 = -6\sin 2x(1 + \cos 2x)^2$, $\frac{d}{dx}\sum_1^\infty \frac{x^n}{n(1+n^2)} = \sum_1^\infty \frac{x^{n-1}}{(1+n^2)}$, $\frac{d}{dx}(\sin 2x)^n = 2n\cos 2x(\sin 2x)^{n-1}$, $\frac{d}{dx}\tan^{-1} 2x = \frac{2}{1+4x^2}$, $\frac{d}{dx}\tan^{-1}\sqrt{x} = \frac{1}{2\sqrt{x}(1+x)}$.

2. $\frac{d}{dx}\left(\frac{c}{s}\right) = \left(s\frac{dc}{dx} - c\frac{ds}{dx}\right)/s^2 = \frac{-s^2-c^2}{s^2}$, so that $\frac{d}{dx}\left(\frac{c}{s}\right) + \left(\frac{c}{s}\right)^2 + 1 = 0$, $\frac{d}{dx}(s.c) = s\frac{dc}{dx} + c\frac{ds}{dx} = s.(-s) + c.c = c^2 - s^2$, $\quad \frac{d}{dx}(c^2 - s^2) = 2c\frac{dc}{dx} - 2s\frac{ds}{dx} = 2c.(-s) - 2s.c = -4s.c$, $\frac{d}{dx}(c^2 + s^2) = 2c\frac{dc}{dx} + 2s\frac{ds}{dx} = 2c.(-s) + 2s.c = 0$.

3. $\frac{d}{dx}f(2x) = 2f'(2x) = 2f(2x)$, since $f' = f$. $\frac{d}{dx}(f(x))^2 = 2f(x).\frac{d}{dx}f(x) = 2f(x).f(x) = 2(f(x))^2$, $\frac{d}{dx}\left(\frac{f(2x)}{(f(x))^2}\right) = \frac{(f(x))^2\frac{d}{dx}f(2x)-f(2x)\frac{d}{dx}(f(x))^2}{(f(x))^4} = \frac{(f(x))^2.2f(2x)-f(2x).2(f(x))^2}{(f(x))^4} = 0$.

4. $\frac{d}{dx}(f^2) = f\frac{df}{dx} + f\frac{df}{dx} = 2f\frac{df}{dx}$, so that $\frac{d}{dx}\left(\frac{1}{f^2}\right) = (f^2\frac{d}{dx}1 - 1\frac{d}{dx}f^2)/f^4 = -2f\frac{df}{dx}/f^4 = -\frac{2}{f^3}\frac{df}{dx}$, by the chain rule with $t = f(x)$, $\frac{d}{dx}\sin f = (\frac{d}{dt}\sin t)\frac{dt}{dx} = \cos t\frac{dt}{dx} = \cos(f)\frac{df}{dx}$, $\frac{d}{dx}\{x^2 f(\frac{1}{x})\} = x^2\frac{d}{dx}f(\frac{1}{x}) + 2xf(\frac{1}{x})$, where $\frac{d}{dx}f(\frac{1}{x}) = -\frac{1}{x^2}f'(\frac{1}{x})$ by the chain rule, giving $\frac{d}{dx}\{x^2 f(\frac{1}{x})\} = 2xf(\frac{1}{x}) - f'(\frac{1}{x})$.

5. The ratio of consecutive terms is $\frac{x^{n+1}}{n+1}/\frac{x^n}{n} = \frac{nx}{n+1}$ which in the limit $n \to \infty$ converges to x.
By the ratio test, we have convergence for $|x| < 1$ and divergence for $|x| > 1$, so the radius of convergence is $R = 1$. For $|x| < 1$, the derivative of the sum of the series is $dg/dx = 1 + x + x^2 + ... = 1/(1 - x)$.

Chapter 6

1. $x^4 = 1 + 4(x - 1) + 6(x - 1)^2 + 4(x - 1)^3 + (x - 1)^4$.

2. (a) $\cos x = 1 - \frac{x^2}{2!} + \frac{x^4}{4!} - \frac{x^6}{6!} + ...$.
   (b) $\cos x = -(x - \frac{\pi}{2}) + \frac{(x-\frac{\pi}{2})^3}{3!} - \frac{(x-\frac{\pi}{2})^5}{5!} + ...$.

3. $f(x) = (1+x)^{-1/2}$, $f'(x) = -\frac{1}{2}(1+x)^{-3/2}$, $f^{(2)}(x) = \frac{1}{2} \cdot \frac{3}{2}(1+x)^{-5/2}$, $f^{(3)}(x) = -\frac{1}{2} \cdot \frac{3}{2} \cdot \frac{5}{2}(1+x)^{-7/2}$, ...

At $x = 0$, $f^{(n)}(0) = \frac{(-1)^n . 1.3....(2n-1)}{2^n}$.

Taylor series $(1+x)^{-1/2} = 1 - \frac{1}{2}x + \frac{1.3}{2.4}x^2 - \frac{1.3.5}{2.4.6}x^3 + ...$ Ratio of x^{n+1} term to x^n term is $-[(2n+1)/2(n+1)]x$ which converges to $-x$ as $n \to \infty$. Convergence for $|x| < 1$, divergence for $|x| > 1$; $R = 1$. If $R > 1$, series converges for $|x| = 1$, which implies convergence at $x = -1$; but $(1+x)^{-1/2}$ is undefined at $x = -1$.

4. $\frac{d}{dx}(c^2 + s^2) = 0$ hence $c^2 + s^2$ is a constant function; since $c^2 + s^2 = 1$ at $x = 0$, we must have $c^2 + s^2 = 1$ for all x. $\frac{d}{dx}(\frac{s}{c}) = (c\frac{ds}{dx} - s\frac{dc}{dx})/c^2 = (c.c - s.(-s))/c^2 = (c^2 + s^2)/c^2 = 1/c^2$. $\frac{d}{dx}(\frac{c}{s}) = (s\frac{dc}{dx} - c\frac{ds}{dx})/s^2 = (s.(-s) - c.c)/s^2 = -(c^2 + s^2)/s^2 = -1/s^2$.

5. $\frac{d}{dx}(2x - x^3) = 2 - 3x^2$; function is increasing for $-\sqrt{\frac{2}{3}} < x < \sqrt{\frac{2}{3}}$, decreasing for $x < -\sqrt{\frac{2}{3}}$ and for $x > \sqrt{\frac{2}{3}}$. For $x > 0$, maximum value is $\frac{4}{3}\sqrt{\frac{2}{3}}$, and occurs at $x = \sqrt{\frac{2}{3}}$.

6.

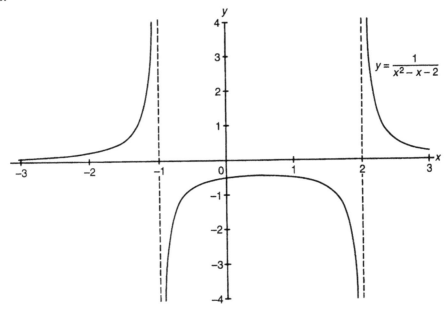

$$y = \frac{1}{x^2 - x - 2}$$

$\frac{dy}{dx} = -\frac{(2x-1)}{(x^2-x-2)^2}$; $\frac{dy}{dx} = 0$ at $x = \frac{1}{2}$, and for $-1 < x < 2$ we have $\frac{dy}{dx} > 0$ for $x < \frac{1}{2}$ and $\frac{dy}{dx} < 0$ for $x > \frac{1}{2}$; hence local maximum at $x = \frac{1}{2}$.

(a) Maximum value is $-\frac{4}{9}$ for $-1 < x < 1$.

(b) Maximum value is $-\frac{1}{2}$ (at $x = 1$) for $1 \leq x < 2$.

7. $\frac{d}{dx}(\frac{2x-3}{2-x^2}) = \frac{2(x^2-3x+2)}{(2-x^2)^2} = 0$ at $x = 1$, $x = 2$. $\frac{dg}{dx} > 0$ for $0 < x < 1$, $\frac{dg}{dx} < 0$ for $1 < x < \sqrt{2}$; $x = 1$ local maximum. $\frac{dg}{dx} < 0$ for $\sqrt{2} < x < 2$, $\frac{dg}{dx} > 0$ for $x > 2$; $x = 2$ local minimum.

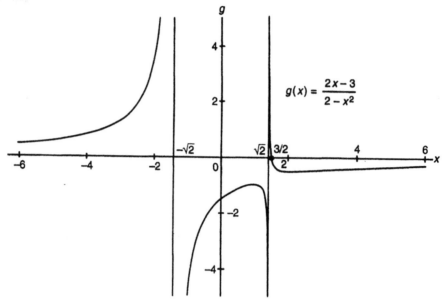

$$g(x) = \frac{2x - 3}{2 - x^2}$$

8. Global minima at $x = -1, 0, 1$; local maxima at $x = \pm\frac{1}{\sqrt{3}}$.

9. (a) $g(x_0) = x_0 - f(x_0)/f'(x_0) = x_0$, since $f(x_0) = 0$.

(b) $g'(x) = 1 - \frac{[(f'(x))^2 - f(x)f''(x)]}{(f'(x))^2} = \frac{f(x)f''(x)}{f'(x)^2} = 0$ at $x = x_0$.

(c) $g''(x) = \frac{(f'(x))^2[f(x)f'''(x) + f'(x)f''(x)] - f(x)f''(x)[2f'(x)f''(x)]}{(f'(x))^4}$, so that $g''(x_0) = f''(x_0)/f'(x_0)$.

From (a) and (b), $\lim_{x_1 \to x_0} \frac{g(x_1) - x_0}{x_1 - x_0} = \lim_{x_1 \to x_0} \frac{g(x_1) - g(x_0)}{x_1 - x_0} = g'(x_0) = 0$ so $x_2 = g(x_1)$ is a better approximation to x_0 since, for x_1 close to x_0, $[g(x_1) - x_0]/(x_1 - x_0)$ will be small, so that $|g(x_1) - x_0|$ will be smaller than $|x_1 - x_0|$.

Suppose that, for $|x_1 - x_0| < m$, say, we have $|g(x_1) - x_0| < \frac{1}{2}m$. Then, as n increases $|g(x_n) - x_0| < (\frac{1}{2})^n.m$, so $|g(x_n) - x_0| \to 0$ as $n \to \infty$, implying $|x_{n+1} - x_0| \to 0$ and $x_n \to x_0$ in the limit.

If $x^3 - x - 1 = 0$, we have $g(x) = x - (x^3 - x - 1)/(3x^2 - 1)$. The root is 1.324718 to six decimal places.

Chapter 7

1. $\frac{d}{dx}\exp(-3x) = -3\exp(-3x)$, $\frac{d}{dx}\exp(-x^2) = -2x\exp(-x^2)$, $\frac{d}{dx}e^{2x}\cos 3x = e^{2x}(2\cos 3x - 3\sin 3x)$, $\frac{d}{dx}x^2 e^{x^2} = 2x(1 + x^2)e^{x^2}$, $\frac{d}{dx}\ln(3x) = \frac{1}{x}$, $\frac{d}{dx}\log_{10}(3x) = \frac{1}{x\ln(10)}$, $\frac{d}{dx}\ln(1 + x^2) = \frac{2x}{1 + x^2}$, $\frac{d}{dx}x^2(\ln(x))^2 = 2x((\ln(x)^2 + \ln(x))$, $\frac{d}{dx}x^x = (1 + \ln(x))x^x$.

2. The derivative of $\ln(1 + x)$ at $x = 0$ is $\lim_{h \to 0}[\ln(1 + h) - \ln(1)]/h$. Since $\ln(1) = 0$, this implies $\lim_{h \to 0}[\ln(1 + h)]/h = 1$. We can take the limit $h \to 0$ by setting $h = 1/x$ and letting x tend to infinity, so that $\lim_{x \to \infty} x\ln(1 + (1/x)) = 1$. Hence $\ln(g(x)) = x\ln(1 + (1/x)) \to 1$ as $x \to \infty$. So $\exp \ln g(x) = (1 + (1/x))^x \to \exp 1 = e$.

3. $\frac{d}{dx}\{\frac{(e^x-e^{-x})^2}{e^{2x}}\} = 4(e^{-2x} - e^{-4x})$, $\exp(-3\ln(x)) = \frac{1}{x^3}$ $(x > 0)$, $\ln(x^2 + x^3) = 2\ln(x) + \ln(1 + x)$.

4. $\frac{d}{dx}(e^x \sin x) = e^x(\sin x + \cos x)$, $\frac{d^2}{dx^2}(e^x \sin x) = 2e^x \cos x$, $\frac{d^3}{dx^3}(e^x \sin x) = 2e^x(\cos x - \sin x)$, $\frac{d^4}{dx^4}(e^x \sin x) = -4e^x \sin x$;   so $\frac{d^{4n}}{dx^{4n}}(e^x \sin x) = (-4)^n e^x \sin x$.

5. Maximum value is $1/2e$.

6. $\cosh x \cosh y + \sinh x \sinh y = \frac{(e^x+e^{-x})(e^y+e^{-y})+(e^x-e^{-x})(e^y-e^{-y})}{4} = \frac{2(e^{x+y}+e^{-x-y})}{4} = \frac{(e^{(x+y)}+e^{-(x+y)})}{2} = \cosh(x + y)$.

7. $(1 + i)^4 = -4$, $(1 + 2i)(1 + 3i) = -5(1 - i)$, $\frac{1}{(1+i)} = \frac{1}{2}(1 - i)$, $\frac{1-i}{2i-1} = -\frac{1}{5}(3 + i)$, $\exp(i\frac{\pi}{2}) = i$.

8. $x = r\cos\theta$, $y = r\sin\theta$; so $re^{i\theta} = r(\cos\theta + i\sin\theta) = x + iy = z$. If $x = y = 1$, then $r = \sqrt{2}$, $\theta = \frac{\pi}{4}$, so $z = \sqrt{2}e^{\frac{i\pi}{4}}$. Hence $(1 + i)^{10} = (\sqrt{2})^{10}e^{\frac{i\pi.10}{4}} = 2^5(\cos\frac{5\pi}{2} + i\sin\frac{5\pi}{2}) = 2^5 i = 32i$.

9. If $z_1 = a_1 + ib_1$ and $z_2 = a_2 + ib_2$, then $\overline{z_1 + z_2} = \overline{a_1 + a_2 + i(b_1 + b_2)} = a_1 + a_2 - i(b_1 + b_2) = a_1 - ib_1 + a_2 - ib_2 = \bar{z}_1 + \bar{z}_2$.
$\overline{z_1 z_2} = \overline{(a_1 + ib_1)(a_2 + ib_2)} = \overline{a_1 a_2 - b_1 b_2 - i(a_1 b_2 + b_1 a_2)} = (a_1 - ib_1)(a_2 - ib_2) = \bar{z}_1 \bar{z}_2$.
$\overline{(\frac{z_1}{z_2})} = \overline{(\frac{z_1 \bar{z}_2}{z_2 \bar{z}_2})} = \overline{(\frac{z_1 \bar{z}_2}{a_2^2 + b_2^2})} = \frac{\bar{z}_1 z_2}{(a_2^2 + b_2^2)} = \frac{\bar{z}_1 z_2}{z_2 \bar{z}_2} = \frac{\bar{z}_1}{\bar{z}_2}$.

10. By the chain rule, $\frac{d}{dx}x^c = \frac{d}{dx}\exp(c\ln(x)) = \exp(c\ln(x))\frac{d}{dx}c\ln(x) = \frac{c}{x}\exp(c\ln(x))$. However, $cx^{c-1} = c\exp((c-1)\ln(x)) = c\exp(-\ln(x)) \times \exp(c\ln(x))$.
Here $\exp(-\ln(x)) = 1/\exp(\ln(x)) = 1/x$, so that $cx^{c-1} = \frac{c}{x}\exp(c\ln(x)) = \frac{d}{dx}x^c$.

11. If $y = \sinh^{-1}x$, $x = \sinh y$. Since $\sinh$ is monotonic increasing, this function is injective and hence has an inverse. $\frac{dx}{dy} = \cosh y$. Since $\cosh^2 y - \sinh^2 y = \cosh^2 y - x^2 = 1$, and $\cosh y > 0$, we have $\cosh y = \sqrt{1 + x^2}$. Hence $\frac{dy}{dx} = \frac{1}{(dx/dy)} = \frac{1}{\sqrt{1+x^2}}$, so that $\frac{d}{dx}\sinh^{-1}x = \frac{1}{\sqrt{1+x^2}}$. The function $\cosh x$ is increasing on the interval $[0, \infty)$ and hence injective on this interval, with range $[1, \infty)$; so the function $\cosh^{-1}x$ is defined, with domain $[1, \infty)$ and range $[0, \infty)$, by $y = \cosh^{-1}x$, $x = \cosh y$. For $x > 1$, we then have $\frac{dx}{dy} = \sinh y = \sqrt{\cosh^2 y - 1} = \sqrt{x^2 - 1}$. Hence $\frac{dy}{dx} = \frac{d}{dx}\cosh^{-1}x = \frac{1}{\sqrt{x^2-1}}$ for $x > 1$.

Chapter 8

1. Respective antiderivatives are $\frac{\sin 4x}{4}$, $\frac{1}{4}\exp(2x^2)$, $\sin x - \frac{2}{3}\sin^3 x + \frac{1}{5}\sin^5 x$, $\frac{1}{16}(\frac{e^{4x}}{4} + 2e^{2x} + 6x - 2e^{-2x} - \frac{e^{-4x}}{4})$.

2. $\int \exp\sqrt{x}\,dx = 2(\sqrt{x}-1)\exp\sqrt{x}+c$, $\int \frac{(\ln(x))^2}{x}\,dx = -\frac{1}{4(1+x^4)}+c$, $\int \frac{(\ln(x))^2}{x}\,dx$
$=\frac{(\ln(x))^3}{3}+c$, $\int \frac{x^4}{1+x}\,dx = \frac{(1+x)^4}{4} - \frac{4(1+x)^3}{3} + 3(1+x)^2 - 4(1+x) + \ln(1+x)+c$,
$\int x^2 \ln(x)\,dx = \frac{1}{9}(3\ln(x)-1)x^3+c$, $\int \sin^{-1}x\,dx = x\sin^{-1}x + \sqrt{1-x^2}+c$.

3. $\int x^2 \cos 2x\,dx = \frac{1}{2}x^2\sin 2x + \frac{1}{2}x\cos 2x - \frac{1}{4}\sin 2x + c$,
$\int x^4 \ln(x)\,dx = \frac{x^5}{5}\ln(x) - \frac{x^5}{25}+c$, $\quad \int \tan^{-1}x\,dx = x\tan^{-1}x - \frac{1}{2}\ln(1+x^2)+c$.
$\int \sqrt{1+x^2}\,dx = x\sqrt{1+x^2} - \int x\frac{d}{dx}\sqrt{1+x^2}\,dx = x\sqrt{1+x^2} - \int \frac{x^2}{\sqrt{1+x^2}}\,dx =$
$x\sqrt{1+x^2} - \int \sqrt{1+x^2}\,dx + \int \frac{1}{\sqrt{1+x^2}}\,dx$.
So $2\int \sqrt{1+x^2}\,dx = x\sqrt{1+x^2} + \int \frac{1}{\sqrt{1+x^2}}\,dx$, giving
$\int \sqrt{1+x^2}\,dx = \frac{1}{2}\{x\sqrt{1+x^2} + \sinh^{-1}x\} + c$.

4. $\int \sqrt{1+x^2}\,dx = \int \sqrt{1+\sinh^2\theta}.\cosh\theta\,d\theta = \int \cosh^2\theta\,d\theta = \int \frac{1+\cosh 2\theta}{2}\,d\theta =$
$\frac{\theta}{2} + \frac{\sinh 2\theta}{4} + c = \frac{\theta + \sinh\theta\,\cosh\theta}{2} + c = \frac{1}{2}\{\sinh^{-1}x + x\sqrt{1+x^2}\} + c$.

5. If $\quad y = \tanh^{-1}x \quad$ then $\quad x = \tanh y = \frac{\sinh y}{\cosh y}$, $\quad$ so $\quad$ that
$\frac{dx}{dy} = \frac{\cosh^2 y - \sinh^2 y}{\cosh^2 y} = 1 - \tanh^2 y = 1 - x^2$. $\quad$ Hence $\quad \frac{dy}{dx} = \frac{d}{dx}\tanh^{-1}x = \frac{1}{1-x^2}$
$(-1 < x < 1)$. $\int \frac{1}{1-x^2}\,dx = \int \frac{1}{2}(\frac{1}{1+x} + \frac{1}{1-x})\,dx = \frac{1}{2}\{\ln(1+x) - \ln(1-x)\} + c$.
Hence $\tanh^{-1}x = \frac{1}{2}\ln(\frac{1+x}{1-x}) + c$ for $-1 < x < 1$, and considering this equation
at $x = 0$, we have $c = 0$.

6. $\frac{d}{dt}\tanh^{-1}(\frac{kv}{c}) = \frac{1}{1-(\frac{kv}{c})^2} \times \frac{k}{c}\frac{dv}{dt} = \frac{kc}{c^2-k^2v^2}\frac{dv}{dt} = kc$. $\quad$ Hence $\quad \tanh^{-1}(\frac{kv}{c}) = kct + a$,
where $\quad a = 0 \quad$ since $\quad v = 0 \quad$ when $\quad t = 0$. $\quad$ So $\quad v = \frac{c}{k}\tanh(kct)$. $\quad$ Since
$\tanh(kct) = \frac{e^{kct}-e^{-kct}}{e^{kct}+e^{-kct}} = \frac{1-e^{-2kct}}{1+e^{-2kct}} \rightarrow 1$ as $t \rightarrow \infty$, we have $v \rightarrow \frac{c}{k}$.

7. $\frac{d}{dt}\tan^{-1}s = \frac{1}{1+s^2}\frac{ds}{dt} = 1$. So $\tan^{-1}s = t + c$. Since $s = 0$ at $t = 0$, we have $c = 0$
and $s = \tan t$.

8. $\frac{d}{dt}((\frac{dz}{dt})^2 + 2gz) = 2\frac{dz}{dt}\frac{d^2z}{dt^2} + 2g\frac{dz}{dt} = 2\frac{dz}{dt}(\frac{d^2z}{dt^2} + g) = 0$. So $(\frac{dz}{dt})^2 + 2gz = c$; since
$\frac{dz}{dt} = u_0$ when $z = 0$, we have $c = u_0^2$ and $(\frac{dz}{dt})^2 = u_0^2 - 2gz$.

9. $\int \frac{1}{(1+x^2)^2}\,dx = \int \frac{1}{\sec^4\theta}\sec^2\theta\,d\theta = \int \cos^2\theta\,d\theta = \int \frac{1+\cos 2\theta}{2}\,d\theta = \frac{\theta}{2} + \frac{\sin 2\theta}{4} + c =$
$\frac{\theta + \sin\theta\,\cos\theta}{2} + c = \frac{1}{2}(\tan^{-1}x + \frac{x}{1+x^2}) + c$.

10. $\int \frac{x}{x^2-1}\,dx = \int \frac{1}{2}\{\frac{1}{x-1} + \frac{1}{x+1}\}\,dx = \frac{1}{2}\ln(x^2-1) + c$,
$\int \frac{1}{x^2(x^2+1)}\,dx = \int \{\frac{1}{x^2} - \frac{1}{(x^2+1)}\}\,dx = -\frac{1}{x} - \tan^{-1}x + c$,
$\int \frac{x-1}{(x+1)(x-2)^2}\,dx = \int \frac{1}{9}\{\frac{2}{x-2} - \frac{2}{x+1} + \frac{3}{(x-2)^2}\}\,dx =$
$\frac{1}{9}\{2\ln(x-2) - 2\ln(x+1) - \frac{3}{(x-2)}\} + c$,
$\int \frac{x^3}{x^2-1}\,dx = \int \{x + \frac{x}{x^2-1}\}\,dx = \frac{x^2}{2} + \frac{1}{2}\ln(x^2-1) + c$.

11. $\int \frac{1}{x(x^2+1)^2}\,dx = \int \{\frac{1}{x} - \frac{x}{x^2+1} - \frac{x}{(x^2+1)^2}\}\,dx = \ln(x) - \frac{1}{2}\ln(x^2+1) + \frac{1}{2(x^2+1)} + c$.

Chapter 9

1. $\int_0^1 \frac{x^2}{(1+x^2)^2}\,dx = [-\frac{x}{2}.\frac{1}{1+x^2}]_0^1 + \int_0^1 \frac{1}{2(1+x^2)}\,dx = [-\frac{x}{2}.\frac{1}{1+x^2} + \frac{1}{2}\tan^{-1}x]_0^1 = (\pi - 2)/8$.
$\int_0^1 \frac{1}{(1+x^2)^2}\,dx = (\pi + 2)/8$. $\quad\quad \int_0^\infty \frac{1}{(1+x^2)^2}\,dx = \frac{\pi}{4}$; $\quad\quad$ the $\quad\quad$ integral $\quad\quad$ is
$\int_0^{\pi/2} \cos^2\theta\,d\theta = [\frac{\theta}{2} + \frac{\sin 2\theta}{4}]_0^{\pi/2}$.

2. The antiderivative is $-e^{-3x}(\cos x + 3\sin x)/10$; $\int_0^\infty e^{-3x}\sin x\,dx = 1/10$.

3. $\int_1^\infty \frac{1}{t(1+t)}dt = [\ln(\frac{t}{1+t})]_1^\infty = \ln(2)$; the integral may also be written as $\int_0^1 \frac{1}{(1+s)}ds = [\ln(1+s)]_0^1 = \ln(2)$.

4. $\int_0^1 \frac{x}{(x+1)(x+2)}dx = \int_0^1 \{\frac{2}{(x+2)} - \frac{1}{(x+1)}\}dx = 2\ln(3) - 3\ln(2) = \ln(\frac{9}{8})$;

$\int_0^1 \frac{x}{(x+1)(x^2+1)}dx = \int_0^1 \{\frac{1}{2(x^2+1)} + \frac{x}{2(x^2+1)} - \frac{1}{2(x+1)}\}dx = \frac{\pi}{8} - \frac{1}{4}\ln(2)$;

$\int_0^\infty \frac{1}{(x+1)(x^2+1)}dx = \frac{1}{2}\int_0^\infty \{\frac{1}{(x+1)} - \frac{(x-1)}{(x^2+1)}\}dx = \frac{1}{2}[\ln(\frac{x+1}{\sqrt{x^2+1}}) + \tan^{-1} x]_0^\infty = \frac{\pi}{4}$.

5. $\int_0^\pi x^2\sin 2x\,dx = [-\frac{x^2\cos 2x}{2} + \frac{x\sin 2x}{2} + \frac{\cos 2x}{4}]_0^\pi = -\frac{\pi^2}{2}$;

$\int_0^\infty x^2 e^{-2x}dx = [-\frac{e^{-2x}}{4}(2x^2 + 2x + 1)]_0^\infty = \frac{1}{4}$;

$\int_0^1 x^2\ln(x)dx = [\frac{x^3}{3}\ln(x) - \frac{x^3}{9}]_0^1 = -\frac{1}{9}$.

6. $\int_0^1 \frac{1}{\sqrt{x}(1+x)}dx < \int_0^1 \frac{1}{\sqrt{x}}dx = [2\sqrt{x}]_0^1$, which is convergent;

$\int_1^\infty \frac{1}{\sqrt{x}(1+x)}dx < \int_1^\infty \frac{1}{x^{3/2}}dx = [-\frac{2}{x^{1/2}}]_1^\infty$, which is convergent.

Hence $\int_0^\infty \frac{1}{\sqrt{x}(1+x)}dx = \int_0^1 \frac{1}{\sqrt{x}(1+x)}dx + \int_1^\infty \frac{1}{\sqrt{x}(1+x)}dx$ is convergent. With the substitution $x = t^2$, the integral becomes $\int_0^\infty \frac{2}{(1+t^2)}dt = \pi$.

$\int_0^\infty \frac{1}{\sqrt{x}(1+\sqrt{x})}dx = \int_0^\infty \frac{2}{1+t}dt$ which diverges, since $2\ln(1+t) \to \infty$ as $t \to \infty$, and the integral is $\lim_{N\to\infty}[2\ln(1+t)]_0^N$.

7. $\frac{d}{dx}\{x\sqrt{x^2-1}\} = x\frac{d}{dx}\sqrt{x^2-1} + \sqrt{x^2-1}\frac{d}{dx}x = \frac{x^2}{\sqrt{x^2-1}} + \sqrt{x^2-1} =$

$\frac{(x^2-1)+1}{\sqrt{x^2-1}} + \sqrt{x^2-1} = 2\sqrt{x^2-1} + \frac{1}{\sqrt{x^2-1}}$. Hence

$\int_1^{\sqrt{2}} \sqrt{x^2-1}\,dx = \int_1^{\sqrt{2}}\{\frac{1}{2}\frac{d}{dx}(x\sqrt{x^2-1}) - \frac{1}{2\sqrt{x^2-1}}\}dx =$

$\frac{1}{2}[x\sqrt{x^2-1} - \cosh^{-1} x]_1^{\sqrt{2}} = \frac{1}{2}(\sqrt{2} - \cosh^{-1}\sqrt{2})$.

8. $\int_{-\pi}^\pi f(t)\,dt = -\int_\pi^{-\pi} f(-s)\,ds = -\int_{-\pi}^\pi f(s)\,ds$ $(= -\int_{-\pi}^\pi f(t)\,dt)$. Hence $\int_{-\pi}^\pi f(t)\,dt = 0$; this comes from cancellation between the areas for $t < 0$ and $t > 0$, respectively, since if one area is above the axis, say for $t < 0$, the area for $t > 0$ will be equal in magnitude but below the axis.

9. $\frac{d}{dx}\int_x^c f(t)\,dt = \frac{d}{dx}\{-\int_c^x f(t)\,dt\} = -f(x)$;

$\frac{d}{dx}\int_{-x}^x f(t)\,dt = \frac{d}{dx}\{\int_{-x}^0 f(t)\,dt + \int_0^x f(t)\,dt\} =$

$\frac{d}{dx}\{\int_0^x f(-s)\,ds + \int_0^x f(t)\,dt\} = f(-x) + f(x)$.

10. $\frac{d^2x}{dt^2} = f(t) = \frac{d}{dt}\int_0^t f(s)\,ds$; hence $\frac{dx}{dt} = \int_0^t f(s)\,ds + c$, where $c = 0$, since $\frac{dx}{dt} = 0$ when $t = 0$.

Also $\frac{d}{dt}\{t\int_0^t f(s)\,ds - \int_0^t sf(s)\,ds\} = tf(t) + \int_0^t f(s)\,ds - tf(t) = \int_0^t f(s)\,ds = \frac{dx}{dt}$. So $x(t) = t\int_0^t f(s)\,ds - \int_0^t sf(s)\,ds + d$, where $d = 0$, since $x = 0$ when $t = 0$. If $f(t) = a\sin\omega t$, $x(t) = t\int_0^t a\sin\omega s\,ds - \int_0^t as\sin\omega s\,ds = \frac{a}{\omega^2}(\omega t - \sin\omega t)$.

Chapter 10

1. (a) $f(x) = x\tan^{-1} x - \frac{1}{2}\ln(1+x^2) + 1$.

(b) $f(x) = \sqrt{1+2x}$.

(c) $f(x) = 3/(3 - x^3)$.

2. $\frac{d^2y}{dx^2} = \frac{d}{dy}(2y^{3/2})\frac{dy}{dx} = 3y^{1/2}\frac{dy}{dx}$; hence $\frac{d^2y}{dx^2} = 3y^{1/2}.2y^{3/2} = 6y^2$; $\frac{d^3y}{dx^3} = 12y\frac{dy}{dx} = 12y.2y^{3/2} = 24y^{5/2}$; $\frac{d^{2n-1}y}{dx^{2n-1}} = (2n)!y^{\frac{2n+1}{2}}$ is easily checked for $n = 1$; assuming the result holds for a given value of n, we have $\frac{d^{2n}y}{dx^{2n}} = (2n)!\frac{d}{dx}y^{\frac{2n+1}{2}} = (2n)!(\frac{2n+1}{2})y^{\frac{2n-1}{2}}\frac{dy}{dx}$, which on substituting $dy/dx = 2y^{3/2}$ gives $d^{2n}y/dx^{2n} = (2n+1)!y^{n+1}$. Similarly, a further differentiation gives $\frac{d^{2n+1}y}{dx^{2n+1}} = (2n+2)!y^{\frac{2n+3}{2}}$, which is the result for $\frac{d^{2n-1}y}{dx^{2n-1}}$, with n replaced by $n+1$. Hence the given formulae hold for $n = 1$, and are true for $n+1$ if they are true for n; so the formulae hold for all positive integer n. Setting $x = 0$, with $y(0) = 1$, we have $d^{2n-1}y/dx^{2n-1} = (2n)!$, and $d^{2n}y/dx^{2n} = (2n+1)!$. Hence, both for k odd and for k even, we have $d^ky/dx^k = (k+1)!$ at $x = 0$. The solution is $y(x) = 1 + 2x + 3x^2 + 4x^3 + \cdots$. Also $\frac{d}{dx}y^{-\frac{1}{2}} = -\frac{1}{2}y^{-\frac{3}{2}}\frac{dy}{dx} = -\frac{1}{2}y^{-\frac{3}{2}} \times 2y^{\frac{3}{2}} = -1$. Hence $y^{-\frac{1}{2}} = -x + c$, and $y(0) = 1$ implies $c = 1$, giving $y(x) = (1-x)^{-2}$; the derivatives of this function at $x = 0$ can again be evaluated to give $d^ky/dx^k = (k+1)!$ and the earlier power series expansion verified.

3. The general solutions are given, respectively, by $y = A - 2x^2$, $y = Ae^{-4x}$, $y = \frac{1}{4}x - \frac{1}{16} + Ae^{-4x}$.

4. $\frac{d}{dx}(e^{4x}y) = e^{4x}(\frac{dy}{dx} + 4y) = e^{4x}p(x)$. So $e^{4x}y = c + \int_0^x e^{4t}p(t)dt$, and $y(0) = 1 \Rightarrow c = 1$. Hence $y = e^{-4x}(1 + \int_0^x e^{4t}p(t)dt)$. Solution of the DE with $p(t) = t^3$ is $y(x) = \frac{x^3}{4} - \frac{3x^2}{16} + \frac{3x}{32} - \frac{3}{128} + \frac{131}{128}e^{-4x}$.

5. $\frac{dy}{dx} = x^2 + \frac{d}{dx}\int_0^x (y(t))^2dt$, so that $y(x) = \frac{x^3}{3} + \int_0^x (y(t))^2dt + c$, where $y(0) = 0$ implies $c = 0$. $y_0(x) = 0$ and $y_{n+1}(x) = \frac{x^3}{3} + \int_0^x (y_n(t))^2dt$. $y_2(x)$ and $y_3(x)$ both give the estimate $y(1) = 0.35$, which is correct to two decimal places. $y = -\frac{1}{z}\frac{dz}{dx}$ implies $\frac{dy}{dx} = \frac{1}{z^2}(\frac{dz}{dx})^2 - \frac{1}{z}\frac{d^2z}{dx^2}$. Since $\frac{dy}{dx} = x^2 + y^2 = x^2 + \frac{1}{z^2}(\frac{dz}{dx})^2$, this leads to $x^2 = -\frac{1}{z}\frac{d^2z}{dx^2}$, or $\frac{d^2z}{dx^2} + x^2z = 0$.

6. $\frac{d^5f}{dx^5} = \int\frac{d^3f}{dx^3} + 3\frac{df}{dx}\frac{d^2f}{dx^2}$; $\frac{d^6f}{dx^6} = \int\frac{d^4f}{dx^4} + 4\frac{df}{dx}\frac{d^3f}{dx^3} + 3(\frac{d^2f}{dx^2})^2$; $\frac{d^7f}{dx^7} = \int\frac{d^5f}{dx^5} + 5\frac{df}{dx}\frac{d^4f}{dx^4} + 10\frac{d^2f}{dx^2}\frac{d^3f}{dx^3}$.

At $x = 0$, $\frac{d^4f}{dx^4} = 1$, $\frac{d^5f}{dx^5} = 0$, $\frac{d^6f}{dx^6} = 4$, $\frac{d^7f}{dx^7} = 0$. The Taylor series is $f(x) = 1 + \frac{x^2}{2} + \frac{x^4}{24} + \frac{x^6}{180} + \cdots$.

Chapter 11

1. $f(x) = x^2 + 1/x^3$.

2. $f(x) = \frac{x}{2} + cxe^{-2x}$.

3. $\frac{dy}{dx} = c + \int_1^x \frac{h(t)}{t^2}dt + \frac{x.h(x)}{x^2}$, so that $x\frac{dy}{dx} - y = cx + x\int_1^x \frac{h(t)}{t^2}dt + h(x) - x[c + \int_1^x \frac{h(t)}{t^2}dt] = h(x)$. $y(1) = c + \int_1^1 \frac{h(t)}{t^2}dt = c$. The solution in the case $h(x) = x$, $c = 1$, is $y = x[1 + \int_1^x \frac{1}{t}dt] = x(1 + \ln(x))$.

4. The respective solutions are given by $f(x) = (e^{-3x} - e^{6x})/9$; $f(x) = Ae^{6x} + Be^{-2x} - \frac{1}{12} - \frac{e^{-3x}}{9}$.

5. $f(x) = e^{-x} \sin x$.

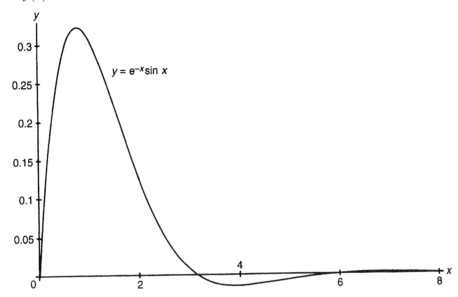

6. $\frac{df}{dx} = \frac{1}{2}e^x \frac{d}{dx} \int_0^x e^{-t} h(t) dt + \frac{1}{2}e^x \int_0^x e^{-t} h(t) dt - \frac{1}{2}e^{-x} \frac{d}{dx} \int_0^x e^t h(t) dt +$
$\frac{1}{2}e^{-x} \int_0^x e^t h(t) dt$, which reduces to $\frac{1}{2}e^x \int_0^x e^{-t} h(t) dt + \frac{1}{2}e^{-x} \int_0^x e^t h(t) dt$ on substituting $\frac{d}{dx} \int_0^x e^{-t} h(t) dt = e^{-x} h(x)$ and $\frac{d}{dx} \int_0^x e^t h(t) dt = e^x h(x)$.

Hence $\frac{d^2f}{dx^2} = \frac{1}{2}e^x \int_0^x e^{-t} h(t) dt - \frac{1}{2}e^{-x} \int_0^x e^t h(t) dt + h(x) = f(x) + h(x)$, so that
$\frac{d^2f}{dx^2} - f = h$. $f(0) = 0$, $\frac{df}{dx} = 0$ at $x = 0$, follow since
$\int_0^0 e^{-t} h(t) dt = \int_0^0 e^t h(t) dt = 0$. The solutions of the given DEs are, respectively,
$f(x) = \frac{1}{2}xe^x - \frac{1}{4}e^x + \frac{1}{4}e^{-x}$;
$f(x) = \frac{1}{4}(x^2 - x)e^x + \frac{1}{8}(e^x - e^{-x})$.

7. $\frac{dy}{dx} = \cos x \int_0^x h(t) \cos t \, dt + \sin x \int_0^x h(t) \sin t \, dt$
$\frac{d^2y}{dx^2} = -\sin x \int_0^x h(t) \cos t \, dt + \cos x \int_0^x h(t) \sin t \, dt + h(x) = -y + h(x)$; hence
$\frac{d^2y}{dx^2} + y = h(x)$.
$y(0) = 0$ and $\frac{dy}{dx} = 0$ at $x = 0$ follow since $\int_0^0 h(t) \cos t \, dt$ and $\int_0^0 h(t) \sin t \, dt$ are
both zero.
The general solution of $\frac{d^2y}{dx^2} + y = \cos x$ is $y = \frac{1}{2}x \sin x + A \cos x + B \sin x$.

8. $c = \pm 2$. The solution is $f(x) = (x^2 - x^{-2})/4$.

9. On differentiating the given equation,
$t\frac{d^3f}{dt^3} + \frac{d^2f}{dt^2} + (t+1)\frac{d^2f}{dt^2} + \frac{df}{dt} + \frac{df}{dt} = 4$, or $t\frac{d^3f}{dt^3} + (t+2)\frac{d^2f}{dt^2} + 2\frac{df}{dt} = 4$. Then
$t\frac{d^4f}{dt^4} + \frac{d^3f}{dt^3} + (t+2)\frac{d^3f}{dt^3} + \frac{d^2f}{dt^2} + 2\frac{d^2f}{dt^2} = 0$, or $t\frac{d^4f}{dt^4} + (t+3)\frac{d^3f}{dt^3} + 3\frac{d^2f}{dt^2} = 0$.
The general equation is $t\frac{d^nf}{dt^n} + (t+n-1)\frac{d^{n-1}f}{dt^{n-1}} + (n-1)\frac{d^{n-2}f}{dt^{n-2}} = 0$, which holds
for $n = 4, 5, 6, \cdots$.
Using the equations to calculate successive derivatives of f at $t = 1$, starting with
$f = 1$, $df/dt = 1$, we have $\frac{d^2f}{dt^2} = 4 - 2\frac{df}{dt} - f = 1$, $\frac{d^3f}{dt^3} = 4 - 3\frac{d^2f}{dt^2} - 2\frac{df}{dt} = -1$, from

which stage the equations are satisfied if successive derivatives alternate in sign. So $d^n f/dt^n = (-1)^n$ for $n \geq 2$. The Taylor series is $f(t) = 1 + (t-1) + \frac{(t-1)^2}{2!} - \frac{(t-1)^3}{3!} + \frac{(t-1)^4}{4!} + \cdots$.

10. $\frac{d}{dt}\{t\frac{df}{dt} + tf\} = t\frac{d^2f}{dt^2} + \frac{df}{dt} + t\frac{df}{dt} + f = t\frac{d^2f}{dt^2} + (t+1)\frac{df}{dt} + f = 4t$.

Hence $t((df/dt) + f) = 2t^2 + c$, where $f = 1$, $df/dt = 1$ at $t = 1$ implies $c = 0$. So $(df/dt) + f = 2t$, for which the general solution is $f(t) = 2(t-1) + Ae^{-t}$. Since $f(1) = 1$, we have $A = e$ and the solution is $f(t) = 2(t-1) + e^{-(t-1)} = 2(t-1) + 1 - (t-1) + \frac{(t-1)^2}{2!} - \cdots = 1 + (t-1) + \frac{(t-1)^2}{2!} - \frac{(t-1)^3}{3!} + \cdots$, which agrees with the Taylor series in exercise 9.

Index